tredition

tredition was established in 2006 by Sandra Latusseck and Soenke Schulz. Based in Hamburg, Germany, tredition offers publishing solutions to authors and publishing houses, combined with worldwide distribution of printed and digital book content. tredition is uniquely positioned to enable authors and publishing houses to create books on their own terms and without conventional manufacturing risks.

For more information please visit: www.tredition.com

TREDITION CLASSICS

This book is part of the TREDITION CLASSICS series. The creators of this series are united by passion for literature and driven by the intention of making all public domain books available in printed format again - worldwide. Most TREDITION CLASSICS titles have been out of print and off the bookstore shelves for decades. At tredition we believe that a great book never goes out of style and that its value is eternal. Several mostly non-profit literature projects provide content to tredition. To support their good work, tredition donates a portion of the proceeds from each sold copy. As a reader of a TREDITION CLASSICS book, you support our mission to save many of the amazing works of world literature from oblivion. See all available books at www.tredition.com.

Project Gutenberg

The content for this book has been graciously provided by Project Gutenberg. Project Gutenberg is a non-profit organization founded by Michael Hart in 1971 at the University of Illinois. The mission of Project Gutenberg is simple: To encourage the creation and distribution of eBooks. Project Gutenberg is the first and largest collection of public domain eBooks.

Tomato Culture: A Practical Treatise on the Tomato

W. W. (William Warner) Tracy

Imprint

This book is part of TREDITION CLASSICS

Author: W. W. (William Warner) Tracy
Cover design: Buchgut, Berlin – Germany

Publisher: tredition GmbH, Hamburg - Germany
ISBN: 978-3-8472-1694-0

www.tredition.com
www.tredition.de

Copyright:
The content of this book is sourced from the public domain.

The intention of the TREDITION CLASSICS series is to make world literature in the public domain available in printed format. Literary enthusiasts and organizations, such as Project Gutenberg, worldwide have scanned and digitally edited the original texts. tredition has subsequently formatted and redesigned the content into a modern reading layout. Therefore, we cannot guarantee the exact reproduction of the original format of a particular historic edition. Please also note that no modifications have been made to the spelling, therefore it may differ from the orthography used today.

TOMATO CULTURE

A PRACTICAL TREATISE ON THE TOMATO, ITS HISTORY, CHARACTERISTICS, PLANTING, FERTILIZATION, CULTIVATION IN FIELD, GARDEN, AND GREENHOUSE, HARVESTING, PACKING, STORING, MARKETING, INSECT ENEMIES AND DISEASES, WITH METHODS OF CONTROL AND REMEDIES, ETC., ETC.

By

WILL W. TRACY

Bureau of Plant Industry, United States Department of Agriculture

ILLUSTRATED

NEW YORK

ORANGE JUDD COMPANY

1907

[Pg ii]

[Pg iii]

To

Dr. F. M. Hexamer

IN HONOR OF HIS LIFELONG EFFORTS FOR THE BETTERMENT OF AMERICAN HORTICULTURAL PRACTICE

Copyright, 1907, by
ORANGE JUDD COMPANY

All rights reserved

WHERE NEW VARIETIES OF TOMATOES ARE DEVELOPED AND TESTED
(By courtesy *American Agriculturist*. Photo by Prof. W. G. Johnson)

PREFACE

This little book has been written in fulfilment of a promise made many years ago. Again and again I have undertaken the work, only to lay it aside because I felt the need of greater experience and wider knowledge. I do not now feel that this deficiency has been by any means fully supplied, but in some directions it has been removed through the kindness of Dr. F. H. Chittenden of the Bureau of Entomology, who wrote the chapter on insect enemies, and of W. A. Orton of the Bureau of Plant Industry, United States Department of Agriculture, who wrote the chapter on diseases of tomatoes.

I have made free use of, without special credit, and am largely indebted to, the writings of Doctor Sturtevant and Professor Goff, Professor Munson of Maine, Professor Halsted of New Jersey, Professor Corbett of Washington, Professor Rolfs of Florida, Professor Bailey of New York, Professor Green of Ohio, and many others. I have also found a vast amount of valuable information in the agricultural press of this country in general. I am also indebted to L. B. Coulter and Prof. W. G. Johnson [Pg vi] for many photographs. My thanks are also due B. F. Williamson, who made the excellent drawings for this book under Professor Johnson's direction.

Tomatoes are among the most generally used and popular vegetables. They are grown not only in gardens, but in large areas in every state from Maine to California and Washington to Florida, and under very different conditions of climate, soil and cultural facilities, as well as of requirements as to character of fruit. The methods which will give the best results under one set of conditions are entirely unsuited to others.

I have tried to give the nature and requirements of the plant and the effect of conditions as seen in my own experience, a knowledge of which may enable the reader to follow the methods most suited to his own conditions and requirements, rather than to recommend the exact methods which have given me the best results.

Will W. Tracy.

TOMATO CULTURE

CHAPTER I

Botany of the Tomato

The common tomato of our gardens belongs to the natural order *Solanaceae* and the genus *Lycopersicum*. The name from *lykos*, a wolf, and *persica*, a peach, is given it because of the supposed aphrodisiacal qualities, and the beauty of the fruit. The genus comprises a few species of South American annual or short-lived perennial, herbaceous, rank-smelling plants in which the many branches are spreading, procumbent, or feebly ascendent and commonly 2 to 6 feet in length, though under some conditions, particularly in the South and in California, they grow much longer. They are covered with resinous viscid secretions and are round, soft, brittle and hairy, when young, but become furrowed, angular, hard and almost woody with enlarged joints, when old. The leaves are irregularly alternate, 5 to 15 inches long, petioled, odd pinnate, with seven to nine short-stemmed leaflets, often with much smaller and stemless ones between them. The larger leaflets are sometimes entire, but more generally notched, cut, or even divided, particularly at the base.

FIG. 2—TOMATO FLOWERS ENLARGED ABOUT 2½ TIMES. SECTION OF FLOWER SHOWN AT RIGHT
(Drawn from a photograph by courtesy of Prof. L. C. Corbett)

The flowers are pendant and borne in more or less branched clusters, located on the stem on the opposite side and usually a little below the leaves; the first cluster on the sixth to twelfth internode from the [Pg 2] ground, with one on each second to sixth succeeding one. The flowers (Fig. 2) are small, consisting of a yellow, deeply five-cleft, wheel-shaped corolla, with a very short tube and broadly lanceolate, recurving petals. The calyx consists of five long linear or lanceolate sepals, which are shorter than the petals at first, but are persistent, and increase in size as the fruits mature. The stamens, five in number, are borne on the throat of the corolla, and consist of long, large anthers, borne on short filaments, loosely joined into a tube and opening by a longitudinal slit on the inside, and this is the chief botanical distinction between [Pg 3] this genus and *Solanum* to which the potato, pepper, night shade and tobacco belong. The anthers in the latter genus open at the tip only. The two genera, however, are closely related and plants belonging to them are readily united by grafting. The Physalis, Husk tomato or Ground cherry is quite distinct, botanically. The pistils of the true tomato are short at first, but the style elongates so as to push the capitate stigma through the tube formed by the anthers, this usually occurring before the anthers open for the discharge of the pollen. The fruit is a two to many-celled berry with central fleshy placenta and many small kidney-shaped seeds which are densely covered with short, stiff hairs, as seen in Figs. 3 and 4.

FIG. 3 – TWO-CELLED TOMATO

FIG. 4 – THREE-CELLED TOMATO

It is comparatively easy to define the genus with which the tomato should be classed botanically, but it is by no means so easy to classify our cultivated varieties into botanical species. We have in cultivation varieties which are known to have originated in gardens and from the same parentage, but which differ [Pg 4] from each other so much in habit of growth, character of leaf and fruit and other respects, that if they had been found growing wild they would unhesitatingly be pronounced different species, and botanists are not agreed as to how our many and very different garden varieties should be classified botanically. Some contend that all of our cultivated sorts are varieties of but two distinct species, while others think they have originated from several.

Classification.—The author suggests the following classification, differing somewhat from that sometimes given, as he believes that the large, deep-sutured fruit of our cultivated varieties and the distinct pear-shaped sorts come from original species rather than from variations of *Lycopersicum cerasiforme*:

Currant tomato, Grape tomato, German or Raisin tomato (*Lycopersicum pimpinellifolium, L. racemiforme*) (Fig. 5).—Universally regarded as a distinct species. Plant strong, growing with many long, slender, weak branches which are not so hairy, viscid, or ill-smelling, and never become so hard or woody as those of the other species. The numerous leaves are very bright green in color, leaflets small, nearly entire, with many small stemless ones between the others. Fruit produced continuously and in great quantity on long racemes like those of the currant, though they are often branched. They continue to elongate and blossom until the fruit at the upper end is fully ripened. Fruit small, less than ½ inch in diameter, spherical, smooth and of a particularly bright, beautiful red color which contrasts well with the bright green leaves, and this abundance of beautifully colored and grace [Pg 5] fully poised fruit makes the plant worthy of more general cultivation as an ornament, though the fruit is of little value for culinary use. This species, when pure, has not varied under cultivation, but it readily crosses with other species and with our garden varieties, and many of these owe their bright red color to the influence of crosses with the above species.

FIG. 5 – CURRANT TOMATO AND CHARACTERISTIC CLUSTERS

[Pg 6]

FIG. 6—RED CHERRY TOMATO

Cherry tomato (*L. cerasiforme*) (Fig. 6).—Plant vigorous, with stout branches which are distinctly trailing in habit. Leaves flat or but slightly curled. Fruit [Pg 7] very abundant, borne in short, branched clusters, globular, perfectly smooth, with no apparent sutures. From ½ to ¾ inch in diameter and either red or yellow in color, two-celled with numerous comparatively small, kidney-shaped seeds. Many of our garden varieties show evidence of crosses with this species, and by many it is regarded as the original wild form of all of our cultivated sorts. These, when they escape from cultivation and become wild plants, as they often do, from New Jersey southward, produce fruit which, in many respects, resembles that of this species in size and form; but they are generally more flattened, globe-shaped, with more or less distinct sutures on the upper side, and I have never seen any fruit of these wild plants which could not be readily distinguished from that of the true Cherry tomato.

Prof. P. H. Rolfs, Director of the Florida experiment station, reports that among the millions of volunteer, or wild, tomatoes he has seen growing in the abandoned tomato fields in Florida, he has never seen a plant with fruit which could not be easily distinguished from that of the true Cherry tomato. Again, one can, by selection and cultivation, easily develop from these wild forms plants producing fruit as large and often practically identical with that of our cultivated varieties, while I have given a true stock of Cherry tomato most careful cultivation on the best of soil for 20 consecutive generations without any increase in size or change in character of the fruit.

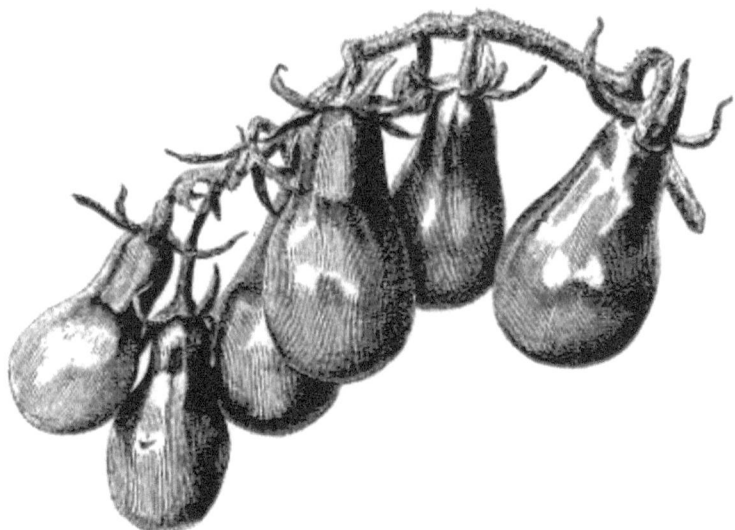

FIG. 7 — PEAR-SHAPED TOMATO

Pear (not Plum) tomato (*L. pyriforme*) (Fig. 7). — Plant exceptionally vigorous, with comparatively few long, stout stems inclined to ascend. Leaves numerous, [Pg 8] broad, flat, with a distinct bluish-green color noticeable, even in the cotyledons. Fruit abundant, borne in short branched or straight clusters of five to ten fruits. It is perfectly smooth, without sutures, and of the shape of a long, slender-necked pear, not over an inch in transverse by 1½ inches in longitudinal diameter. When the stock is pure the fruit retains this form very persistently. The production of egg-shaped or other forms is a sure indication of impure stock. They are bright red, dark yellow, or light yellowish white in color, two-celled, with very distinct central placenta and comparatively few and large seeds. The fruit is inclined to ripen unevenly, the neck remaining green when the rest of the fruit is quite ripe. It is less juicy than that of most of our garden sorts but of a mild and pleasant flavor. This is considered, by many, to [Pg 10] be simply a garden variety, but I am inclined to the belief that it is a distinct species and that the contrary view comes from the study of the impure and crossed stocks resulting from crosses between the true Pear tomato and garden sorts which are frequently sold by seedsmen as pear-shaped. Many garden sorts—like the Plum (Fig. 8), the Egg, the Golden Nugget, Vick's Criterion, etc.—are known to have originated from crosses of the Pear and I think that most, if not all, the garden sorts in which the longitudinal diameter of the fruit is greater than its transverse diameter owe this form to crosses with *L. pyriforme*.

FIG. 8—YELLOW PLUM TOMATO, SHOWING MOST USUAL FORM OF CLUSTER

Cultivated varieties (*L. esculentum*).—This is commonly used as the botanical name of our cultivated varieties, rather than as the name of a distinct species. In western South America, however, there is found growing a wild plant of Lycopersicum which differs from the other recognized species in being more compact in growth, with fewer branches and larger leaves, and carrying an immense burden of fruit borne in large clusters. The fruit is larger than that of the other species but much smaller than that of our cultivated sorts; is very irregular in shape, always with distinct sutures, and often deeply corrugated and bright red in color. The walls are thin; the flesh is soft, with a distinct sharp, acid flavor much less agreeable than that of our cultivated forms of garden tomatoes.

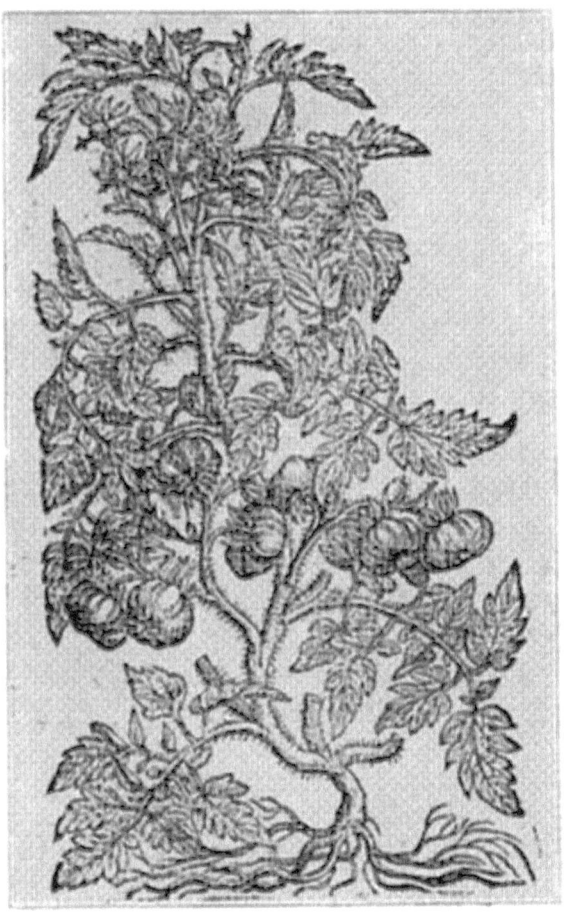

[Pg 11]

FIG. 9 — ONE OF THE FIRST ILLUSTRATIONS OF THE TOMATO
Poma amoris, (*Pomum aureum*), (*Lycopersicum*), 1581

[Pg 12]
FIG. 10 — AN EARLY ILLUSTRATION OF THE TOMATO
(From Morrison's "Historia Universalis," 1680)

This has commonly been regarded by botanists as a degenerate form of our garden tomatoes, rather than as an original species, but I find that, like *L. cerasiforme* and *L. pyriforme*, it is quite fixed under cultivation, except as crossed with other species or with our gar [Pg

13] den varieties, and I believe it to be the original species from which our cultured sorts have been developed, by crossing and selection. Such crosses probably were made either naturally or by natives before the tomato was discovered by Europeans. The earliest prints we have of the tomato (Figs. 9 and 10) are far more like the fruit of this plant than that of *L. cerasiforme*, and the prints of many of the earliest garden varieties and of some sorts which are still cultivated in southern Europe, for use in soups, are like it not only in size and form, but in flavor. These facts make it seem far more probable that our cultivated sorts have come, by crossing, between this and other species rather than by simple development from *L. cerasiforme*.

Prof. E. S. Goff, of Wisconsin, who has made a most careful study of the tomato, expressed the same opinion, writing that it seemed to him that our cultivated sorts must have come from the crossing of a small, round, smooth, sutureless type, with a larger, deep-sutured, corrugated fruit, like that of the Mammoth Chihuahua, but smaller. However this may be, I think that it is wise to throw all of our cultivated garden sorts, except the Pear, the Cherry, and the Grape—which I regard as distinct species—together under the name of *L. esculentum*, even when we know they have originated by direct crosses with the other species; and it is well to classify the upright growing sorts under the varietal names, *L. validum*, and the larger, heavier sorts, as *L. grandifolium*, as has been done by Bailey. (Cyclopedia of Horticulture.)

[Pg 14]

CHAPTER II

History

The garden vegetable known in this country as tomato and generally as tomate in continental Europe, is also known as Wolf-peach and Love Apple in England and America, and Liebesapfel in Germany, Pomme d'Amour in France, Pomo d'oro in Italy, Pomidor in Poland.

Origin of name.—The name tomato is of South American origin, and is derived from the Aztec word *xitomate*, or *zitotomate*, which is given the fruit of both the Common tomato and that of the Husk or Strawberry tomato or Physalis. Both vegetables were highly prized and extensively cultivated by the natives long before the discovery of the country by Europeans, and there is little doubt that many of the plants first seen and described by Europeans as wild species were really garden varieties originated with the native Americans by the variation or crossing of the original wild species.

Different types now common, according to Sturtevant, have become known to, and been described by Europeans in about the following order:

1. Large yellow, described by Matthiolus in 1554 and called Golden apple.
2. Large red, described by Matthiolus in 1554 and called Love apple. [Pg 15]
3. Purple red, described by D'el Obel in 1570.
4. White-fleshed, described by Dodoens in 1586.
5. Red cherry, described by Bauhin in 1620.
6. Yellow cherry, described by Bauhin in 1620.
7. Ochre yellow, described by Bauhin in 1651.
8. Striped, blotched or visi-colored, described by Bauhin in 1651.
9. Pale red, described by Tournefort in 1700.
10. Large smooth, or ribless red, described by Tournefort in 1700.
11. Bronzed-leaved, described by Blacknell in 1750.
12. Deep orange, described by Bryant in 1783.

13. Pear-shaped, described by Dunal in 1805.
14. Tree tomato, described by Vilmorin in 1855.
15. Broad-leaved, introduced about 1860.

The special description of No. 10 by Tournefort in 1700 would indicate that large smooth sorts, like Livingston's Stone, were in existence fully 200 years ago, instead of being modern improvements, as is sometimes claimed; and a careful study of old descriptions and cuts and comparing them with the best examples of modern varieties led Doctor Sturtevant in 1889 to express the opinion that they had fruit as large and smooth as those we now grow, before the tomato came into general use in America, and possibly before the fruit was generally known to Europeans. Even the production of fine fruit under glass is not so modern as many suppose. In transactions of the London Horticultural Society for 1820, John Wilmot is reported to have cultivated under glass in 1818 some 600 plants and gathered from his entire plantings under [Pg 16] glass and in borders some 130 bushels of ripe fruit. It is stated that the growth that year exceeded the demand, and that the fruit obtained was of extraordinary size, some exceeding 12 inches in circumference and weighing 12 ounces each. Thomas Meehan states in *Gardeners' Monthly* for February, 1880, that on January 8, of that year, he saw growing in the greenhouses on Senator Cannon's place near Harrisburg, Pa., at least 1 bushel of ripe fruits, none of which were less than 10 inches in circumference,—a showing which compares with the best to be seen to-day.

Throughout southern Europe the value of the fruit for use in soups and as a salad seems to have been at once recognized, and it came into quite general use, especially in Spain and Italy, during the 17th century; but in northern Europe and England, though the plant was grown in botanical gardens and in a few private places as a curiosity and for the beauty of its fruit, this was seldom eaten, being commonly regarded as unhealthy and even poisonous, and on this account, and probably because of its supposed aphrodisiacal qualities, it did not come into general use in those northern countries until early in the 19th century.

First mention in America, I find of its being grown for culinary use, was in Virginia in 1781. In 1788 a Frenchman in Philadelphia made most earnest efforts to get people to use the fruit, but with little success, and similar efforts by an Italian in Salem, Mass., in 1802, were no more successful. The first record I can find of the fruit being regularly quoted in the market was in New Orleans in 1812, and the earliest records I have been able to find of the seed being [Pg 17] offered by seedsmen, as that of an edible vegetable, was by Gardener and Hipburn in 1818, and by Landreth in 1820. Buist's "Kitchen Gardener" says: "In 1828-9 it (the tomato) was almost detested and commonly considered poisonous. Ten years later every variety of pill and panacea was 'extract of tomatoes,' and now (1847) almost as much ground is devoted to its culture as to the cabbage." In 1834 Professor Dunglison, of the University of Virginia, said: "The tomato may be looked upon as one of the most wholesome and valuable esculents of the garden."

Yet, though the fruit has always received similar commendation from medical men, there has been constant recurring superstition that it is unhealthy. Only a few years ago there was in general circulation a statement that an eminent physician had discovered that eating tomatoes tended to develop cancer. This has been definitely traced to the playful question, asked as a joke by Dr. Dio Lewis, "Didn't you know that eating bright red tomatoes caused cancer?" In more recent years an equally unfounded claim has been made that tomato seeds were responsible for many cases of appendicitis and that it was consequently dangerous to eat the fruit.

I give some quotations for tomatoes in Quincy Hall Market, Boston, with some for other vegetables, for comparison. The records show that during the week ending July 22, 1835, tomatoes were quoted at 50 cents per dozen, cabbage at 50 cents per dozen. For the week ending September 22, 1835, tomatoes were quoted at 25 cents per peck, lima beans, 12½ cents per quart shelled, with comment that tomatoes are in [Pg 18] much demand and a far greater quantity has been sold than in previous years. During the week ending July 22, 1837, tomatoes were quoted at 25 and 50 cents per peck, and the note that they are of good size and were well ripened and came from gardens in the vicinity would indicate that they had at that time early maturing varieties and knew how to grow them. From about 1835 till the present time the cultivation and use of tomatoes have constantly increased both in this country and in Europe, so that now they are one of the most largely grown of our garden vegetables.

A suggestion as to the extent they are now grown in America is the fact that a single seed grower saved in 1903 over 20,000 pounds of tomato seed—an amount sufficient to furnish plants for from 80,000 to 320,000 acres, according to the care used in raising them, the larger quantity not requiring more care than the best growers commonly use. A careful estimate made by the *American Grocer* shows that in 1903 the packing of tomatoes by canners in the United States amounted to 246,775,426 three-pound cans. In addition to the canned tomato, between 200,000 and 250,000 barrels of catsup stock is put up annually, requiring the product of at least 20,000 acres.

It is probable that the area required to produce the fruit that is used fresh at least equals that devoted to the production for preserving, which give us from 400,000 to 500,000 acres devoted to this crop each year in America alone. The fruit is perhaps in more general use in America than elsewhere, but its cultivation and use have increased rapidly in other coun [Pg 19] tries, particularly with the English speaking races. Large quantities are grown in Australia, and immense and constantly increasing quantities are grown under glass in England and adjacent islands, while *The Gardeners' Chronicle* states that in 1903 between 600,000 and 800,000 pounds of fresh fruit were imported into England from other countries.

[Pg 20]

CHAPTER III

General Characteristics of the Plant

In the native home of the tomato, in South America, the conditions of the soil, both as regards composition and mechanical condition, of the moisture both in soil and air, and those of temperature and sunlight, are throughout the growing season not only very favorable for rapid growth, but are uniformly and constantly so. Under such conditions there has been developed a plant which, while vigorous, tenacious of life, capable of rapid growth and enormously productive, is not at all hardy in the sense of ability to endure untoward conditions either in the character of soil, of water supply, or of temperature. A check in the development because of any unfavorable condition is never fully recovered from, but will inevitably affect the total quantity and quality of the fruit produced, even if subsequent favorable conditions result in the rapid and vigorous growth of the plant.

I know of an instance where two adjoining fields belonging to A and B were set with tomatoes, using plants started in the same hotbed from the same lot of seed. The soil was of equal natural fertility and each field received about the same quantity of manure, though that given A's was all well decomposed and worked into the soil, while that given B's was fresh and raw and simply plowed in. A's field was [Pg 21] put into the best possible tilth before setting the plants, and the management of the plants and their cultivation were such as to secure unchecked growth from the time they were pricked out into cold-frames and set in the field until the crop was matured. As long as the plants would permit, the soil was cultivated every few days and kept in a state of perfect tilth.

B's field when the plants were set out was a mass of clods, as it had been plowed, when wet, some time before and never harrowed but once. The plants had been crowded forward as rapidly as possible in the cold-frame, and when set in the field were much higher than A's, but so soft that they were badly checked in transplanting and a great many of them died and had to be reset. The field received but one or two cultivations during the entire season. The growth of the plants in B's field was irregular and uneven instead of steady and uniform as in A's, and though some of the fruits were quite as large, they were not as uniform as A's while the yield per acre was not more than half as much nor the fruit of as good general quality. B had difficulty in disposing of his crop and often had to sell below the market, while A had no trouble in disposing of his at the highest prices for the day. B's crop was a financial loss, while A's returned a most satisfactory profit.

The key to the most successful culture of the tomato is the securing, from the start to finish, of an unchecked uniform growth, though it need not necessarily be a rapid one. The failure to do this is, in my opinion, the principal reason for the comparatively small yield usually obtained, which is very [Pg 22] much less than it would be with better cultural management. The tomato under conditions which I have repeatedly found it practicable to secure, not only in small plantings but in large fields, has proved capable of producing from 1,000 to 1,200 or even more bushels to the acre, and the possible yield per plant is enormous.

As early as 1818 the Royal Horticultural Society of London reports the obtaining of over 40 pounds of fruit of marketable character from a single vine. An acre of such plants would give a yield of over 1,800 bushels of fruit, and many similar yields, and even greater ones, have been recorded for single plants. The yield commonly obtained, even in favorable locations, and by men who have grown tomatoes all their lives, is more often less than 200 bushels to the acre than more. The way to secure a better yield is to study carefully the nature and requirements of the plants and the adaptation of our cultural practice to them.

Life habit of the plant.—The tomato could be described as a short-lived perennial, but its span of life is somewhat variable. Under favorable conditions it will develop from starting seed to first ripe fruit in from 85 to 120 days of full sunshine with a constant day temperature of from 75 to 90° F., and with one from 15 to 20° F. lower at night. The plants will ordinarily continue in full fruit for about 50 to 60 days, after which they generally become so exhausted by excessive production of fruit and the effects of diseases to which they are usually subject that their root action and sap circulation become weaker and [Pg 23] weaker until they die from starvation.

From Philadelphia southward gardeners expect that spring set plants will thus exhaust themselves and die by late summer, and they sow seed in late spring or early summer for plants on which they depend for late summer and fall crops.

Under some conditions, particularly in the Gulf states and in California, tomato plants will not only grow to a much greater size than normal, but will continue to thrive and bear fruit for a longer time. Such a plant grown in Pasadena, Cal., was said to have been in constant bearing for over 10 months. Again, sometimes plants that have produced a full crop of fruits will start new sets of roots and leaves and produce a second and even a third crop, each, however, being produced on new branches and as a result of a fresh set of roots, those which produced the preceding crop having died and disappeared. The period of development, 85 to 120 days of full sunshine at a temperature above 75° F., has been given. The full sunshine and high temperature are essential to such rapid development, and in so far as there is a lack of sunshine from clouds or shade, or the day temperature falls below 75° F. the period will be lengthened, so that in the greater part of the United States the elapsed time between starting seed to ripened fruit is usually as much as from 120 to 150 days and often even longer.

Characteristics of the root.—The roots of the tomato plant, while abundant in number, are short and can only gather food and water from a limited area. A plant of garden bean, for instance, is not [Pg 24] more than half the size of one of the tomato, but its roots extend through the soil to a greater distance, gather plant food from a greater bulk of soil, seem better able to search out and gather the particular food element which the plant needs than do those of the tomato. This characteristic of the latter plant makes the composition of the soil as to the proportion of easily available food elements of great importance. Tomato roots are also exceedingly tender and incapable of penetrating a hard and compact soil, so that the condition of the soil as to tilth is of greater importance with regard to tomatoes than with most garden vegetables.

Another characteristic of the tomato roots is that the period of their active life is short. When young they are capable of transmitting water and nutritive material very rapidly, but they soon become clogged and inefficient to such an extent as to result in the starvation and death of the plant. If the branches of such an exhausted plant be bent over and covered with earth they will frequently start new roots and produce a fresh crop of fruit, or if plants which have made a crop in the greenhouse be transplanted to the garden and cut back, a new set of roots will often develop and the plant will produce a second crop of fruit which, in amount, often equals or exceeds the first one. But such growths come only from new roots springing from the stem—never from an extension of the old root system.

Characteristics of the stem and leaves.—The growth of the stem, and leaves of the young tomato plant is very rapid and, the cellular structure coarse, [Pg 25] loose and open. A young branch is easily broken and when this is done it shows scarcely any fibrous structure—simply a mass of coarse cellular matter which while capable, when young, of transmitting nutritive matter rapidly, soon becomes dogged and inert. This structure not only makes the active life of the leaves short, like that of the roots, but necessitates a fresh growth in order to continue the fruitfulness of the plant and renders the leaves very susceptible to injury from bacterial and fungous diseases. The rapid growth also necessitates an abundance of sunlight.

Characteristics of the blossom.—The inflorescence of the tomato is usually abundant and it is rare that a plant does not produce sufficient blooms for a full crop. The flowers are perfect as far as parts are concerned (Fig. 2) and in bright, sunny weather there is an abundance of pollen, but sunlight and warmth are essential to its maturing into a condition in which it can easily reach the stigma. The structure and development of the flower are such that while occasionally, particularly in healthy plants out of doors, the stigma becomes receptive and takes the pollen as it is pushed out through the stamen tube by the elongating style, it is more often pushed beyond them before the pollen matures, so that the pollen has to reach the stigma through some other means. Usually this is accomplished by the wind, either directly or through the motion of the plants.

Under glass it is generally necessary to assist the fertilization either directly by application or by motion of the plant, this latter only being effective in the middle of a bright sunny day. In the open ground [Pg 26] in cold, damp weather the flowers often fail of fertilization, in which case they drop, and this is often the first indication of a failing of the crop on large, strong vines. I have known of many cases where the yield of fruit from large and seemingly very healthy vines was very light because continual rains prevented the pollenization of the flowers. Such failures, however, do not always come from a want of pollen but may result from an over or irregular supply of water either at the root or in the air, imperfectly balanced food supply, a sapping of the vitality of the plants when young, or from other causes. Insects rarely visit tomato flowers and are seldom the means of their fertilization.

Characteristics of the fruit.—The fruit of the original species from which our cultivated tomatoes have developed was doubtless a comparatively small two to many-celled berry, with comparatively dry central placenta and thin walls. In some species the cells were indicated by distinct sutures, forming a rough or corrugated fruit. It has improved under cultivation by increase in size, the material thickening of the cell walls, the development of greater juiciness and richer flavor and a decrease in the size and dryness of the placenta, as well as the breaking up of the cells by fleshy partitions resulting in the disappearance of the deep sutures and an improvement in the smoothness and beauty of the fruit. (Fig. 11.)

The quality of the fruit is largely dependent upon varietal differences, to be spoken of later, but it is also influenced by conditions of growth—such as the proportion of the nutritive elements found in the soil, [Pg 27] the proper supply of moisture, the degree and uniformity of temperature and, most of all, the amount of sunlight. Sudden changes of temperature and moisture often result in cracks and fissures in the skin and flesh, which not only injure the appearance but affect the flavor of the fruit.

FIG. 11 – TYPICAL BUNCH OF MODERN TOMATOES
Contrast with Figs. 9 and 10

[Pg 28]

CHAPTER IV

Essentials for Development

Sunlight.—Abundant and unobstructed sunlight is the most essential condition for the healthy growth of the tomato. It is a native of the sunny South and will not thrive except in full and abundant sunlight. I have never been able to grow good tomatoes in the shade even where it is only partial. The entire plant needs the sunlight. The blossoms often fail to set and the fruit is lacking in flavor because of shade, from excessive leaf growth, or other obstruction.

The great difficulty in winter forcing tomatoes under glass in the North comes from the want of sunlight during the short days of the winter months. Were it not for the short winter days of the higher latitudes limiting the hours of sunshine, tomatoes could be grown under glass in the northern states to compete in price, when the better quality of vine-ripened fruits is considered, with those from the Gulf states. Growers are learning that tomatoes can be profitably grown under glass during the longer spring days, and consumers are beginning to appreciate the superior quality of fruit ripened on the vine over that picked green and ripened in transit. At no time is this need of abundance of light of greater importance than when the plants are young and, if they fail to receive it, no subsequent favorable conditions will enable them to recover fully from its ill effects. It is not so much [Pg 29] the want of room for the roots as of light for the leaves that makes the plants which have been crowded in the seed-beds so weak and unprofitable.

I once divided 100 young tomato plants, about 2 inches high, into four lots of 25 each, numbering them 1, 2, 3 and 4. The plants of lots No. 1 and 2 were set equal distance apart in box A, and those of lots No. 3 and 4 in the same way in box B; both boxes being about 16 inches wide, 40 inches long and 4 inches deep. The two boxes were set together across the side bench of a greenhouse with the outer edge against a board wall some 2½ feet high, so that the plants at the end of the box near the wall received much less light than those at the other end. They remained there about five weeks and then were taken out and the plants set in the open ground. During the five weeks box A, containing lots No. 1 and 2, was changed, end for end, every day so that those two lots of plants received nearly an equal amount of sunlight, but box B was not changed so that lot No. 3, at one end of the box, was constantly near the walk and in the full light, while lot No. 4, at the other end of the box, was constantly near the wall and in partial shade. The effect on the growth of the plants was very marked. The plants of lot No. 4 were nearly twice as high, but with much softer stems and leaves than those of lot No. 3. The plants received equal care when set side by side in the open ground and at the time the first fruit was gathered seemed of equal size and vigor, but the total yield of fruit of lots No. 1, 2 and 3 was very nearly the same and in each case at the rate of over 100 bushels an acre more [Pg 30] than that from lot No. 4. This is but one of the scores of experiences which have led me to appreciate, in some degree, the necessity of plenty of sunlight for the best development of the tomato.

Heat.—The plant thrives best out of doors in a dry temperature of 75 to 85° F., or even up to 95° F., if the air is not too dry and is in gentle circulation. The rate of growth diminishes as the temperature falls below 75° until at 50° there is practically no growth; the plant is simply living at a poor dying rate and if the growth, particularly in young plants, is checked in this way for any considerable time they will never produce a full crop of fruit, even if the plants reach full size and are seemingly vigorous and healthy. The plant is generally killed by exposure for even a short time to freezing temperature, though young volunteer plants in the spring are frequently so hardened by exposure that they will survive a frost that crusts the ground they stand in; but such exposure affects the productiveness of the plant, even if it subsequently makes a seemingly vigorous and healthy growth. Under glass, plants usually do best in a temperature somewhat lower than is most desirable out of doors. I think this is due to the inevitable obstruction of the sunlight and the lack of perfect ventilation.

Moisture.—Although the tomato is not a desert plant and needs a plentiful supply of water, it suffers far more frequently, particularly when the plants are young, from an over-supply than from the want of water. Good drainage at the root and warm, dry, sunny air, in gentle motion, are what it delights in. [Pg 31] Good drainage is essential not only to the best growth of the plant but to the production of any fruit of good quality. So important is this feature that though it can be readily proved that, other things being equal, the tomato will give larger yield and better fruit on well drained clay loam than on sandy soil, yet it is more generally and more successfully planted on sandy lands simply because they are usually better drained and on this account give better crops. While excess of water in the soil is most injurious to the young and growing plant, an abundance of it at the time the fruit swells and ripens is very essential, and a want of it at that time results in small and imperfect fruit of poor flavor. Excessive moisture in the air is just as injurious as at the root. In my personal experience I have known of more failures in tomato crops, at least in the northern states, to come from a season of persistent rains and damp atmosphere at the time when the plants should be in bloom and setting fruit than from any other climatic cause.

Food supply.—The tomato is not a gross feeder nor is the crop an exhaustive one, but the plant is very particular as to its food supply. It is an epicure among plants and demands that its food shall not only be to its taste in quality but that it be well served. In order for the plant to do its best, or even well, it is essential that the food elements be in the right proportions and readily available. If there is a deficiency of any single element there will be but a meager crop of fruit, no matter how abundant the supply of the others. An oversupply of an element, especially nitrogen, is hardly less injurious and will actually les [Pg 32] sen the yield of fruit though it may increase the size of the vine. Not only must the food be in right proportions but in such condition as to be readily available. Tomato roots have little power to wrest plant food from the soil. The use of coarse, unfermented manure is even more unsatisfactory with this than with other crops. The enormous yields sometimes obtained by English gardeners from plants grown under glass result from a supply of food of the right proportions and in solution, instead of incorporating it in a crude condition with the soil.

Cultivation.—The tomato is grown in all parts of the United States and under very different conditions, not only as to climate and soil but as to the facilities for growing and handling the crop and the way in which it is done. What would be ideal conditions of soil and the most advantageous methods under some conditions would not be at all desirable in others. In some cases the largest possible yield an acre, in others fruit at the lowest cost a bushel, or at the earliest possible date, or in a continuous supply and of the best quality, is the greatest desideratum. It is impossible to give specific instructions which would be applicable to all these varying conditions and requirements; so I give general cultural directions for maximum crops with variations suggested for special conditions and requirements, and then the reader may follow those which seem best suited to his individual conditions.

[Pg 33]

CHAPTER V

Selection of Soil for Maximum Crop

Large yields of tomatoes have been, and can be, obtained from soils of varying composition, from a gumbo prairie, a black marsh muck, or a stiff, tenacious clay, to one of light drifting sand, provided other conditions, such as drainage, tilth and fertility are favorable. The Connecticut experiment station and others have secured good results from plants grown under glass in a soil of sifted coal ashes and muck, or even from coal ashes alone, the requisite plant food being supplied in solution. But a maximum crop could never, and a full one very seldom, be produced on a soil, no matter what its composition, which could not be, or was not put into and kept in a good state of tilth, or on one which was poorly drained, sodden or sour, or which was so leachy that it was impossible to retain a fair supply of moisture and of plant food.

Of the 10 largest yields of which I have personal knowledge and which ran from 1,000 to 1,200 bushels of fruit (acceptable for canning and at least two-thirds of it of prime market quality) an acre, four were grown on soils classed as clay loam, two on heavy clay—one of which was so heavy that clay for making brick was subsequently taken from the very spot which yielded the most and best fruit—one on what had been a black ash swamp, one on a sandy muck, two on a sandy loam and one on a light sand [Pg 34] made very rich by heavy, annual manuring for several years. They were all perfectly watered and drained, in good heart, liberally fertilized with manures of proved right proportions for each field, and above all, the fields were put into and kept in perfect tilth by methods suited to each case; while the plants used were of good stock and so grown, set and cultivated that their growth was never stopped or hardly checked for even a day. These conditions as to soil and culture, together with seasons of exceptionally favorable weather, resulted in uniformly large crops on these widely different soils.

FIG. 12—TOMATOES TRAINED TO STAKES ON A GEORGIA FARM

The composition of the soil, then, as to its proportions of sand or clay is of minor importance as regards a maximum yield or as to quality of the fruit, except as it affects our ability to put and keep the soil in good physical condition. The tomato crop, however, particularly when the plants are trimmed and trained to stakes, as is the usual practice in the South, as seen in Fig. 12, with crops grown for early shipment, necessitates in the trimming and training of the plants and the gathering of the fruit when it is in the right degree of maturity for shipment a great deal of trampling of the surface regardless of whether it is wet or dry. Consequently if the surface soil has any considerable proportion of clay there is danger of compacting and even puddling it by working when wet, to the great detriment of the crop. Again, a more or less sandy surface soil can be much more easily worked than one with a large proportion of clay. For these reasons our choice of a soil for the lowest cost a bushel and probably for a maximum yield should be a rich sandy [Pg 36] or sandy loam surface soil overlying a well-drained clay sub-soil. I would prefer one which was originally covered with a heavy growth of beech and maple timber, though I should want it to be "old land" at the time. Tomatoes do not succeed as well on prairie soils, particularly if they are at all heavy, as they do on timbered lands, but one need not despair of a profitable crop of tomatoes on any soil which would give a fair crop of corn or of cotton.

For early-ripening fruit.—Sometimes the profit and satisfaction from a tomato crop depend more largely upon the earliness of ripening than upon the amount of yield or cost of growing. In such cases a warm, sandy loam, or even a distinctly sandy soil, is to be preferred, as this is apt to be warmer and the fruit will be matured much earlier on it than on a heavier soil. It is essential, however, that it be well drained and warm. Often lands classed as sandy are really colder than some of those classed as clay, and such soils should be carefully avoided if early maturity is important.

For the home garden.—Here we seldom have a choice, but no one need despair and abandon effort, no matter what the soil may be, for it is quite possible to raise an abundant home supply on any soil and that, too, without inordinate cost and labor. Some of the most prolific plants and the finest fruits I have ever seen were grown in a village lot which five years before had been filled in to a depth of 3 to 10 feet with clay, coal ashes and refuse from a brick and coal yard. In another instance magnificent fruit was grown in a garden where the soil was originally made up chiefly [Pg 37] of sawdust mixed with sand, drawn on a foundation of sawmill edgings so as to raise it above the water of a swamp. Where one has to contend with such conditions he should make an effort to create a friable soil with a supply of humus by adding the material needed. A very few loads, sometimes even a single load, of clay or sand will greatly change the character of the soil of a sufficient area to grow the one or two dozen plants necessary for a family supply. In the two cases mentioned, the owner of the first named garden used both sand and sawdust to lighten his soil, while the second drew a great many loads of clay on his.

Growing under glass.—I would make up a soil composed of about three parts rotted sod, two or three parts of well-rotted stable manure (and it is very important that it be well decomposed) and one part either of coarse, sharp sand, sandy loam or clay loam, according as the sod soil is light or heavy, the aim being to form a rich, light, open soil rather than one which is as heavy and compact as desirable for some plants. If sod soil is not available, of course, garden loam can be substituted, but it is very important that the soil be thoroughly mixed, and desirable that it be prepared sometime before it is to be used. Some growers use the same soil for several crops, simply adding some fresh manure; but, if so used, it is important that it be stirred and thoroughly re-mixed and sterilized.

[Pg 38]

CHAPTER VI

Exposure and Location

In sections where there is danger of the plants being killed by early fall frosts before they have ripened their entire crop, exposure of the field is sometimes of importance in determining the marketable yield.

A gentle inclination to the south, with a protection of higher land or timber on the sides from which frost or high winds are most likely to come, is the best. A steep descent to the south, shut in by high land to the east and west, so as to form a hot pocket, is not favorable for a maximum crop although it may give a smaller yield of early ripening fruit; nor is a small field entirely surrounded by forest desirable.

I once knew of a field, of about two acres, sloping to the south and entirely surrounded by heavy timber, on which two or three tomato crops were failures when other fields on the same farm gave large yields, but after the timber on the south and east had been cut away this field generally gave the largest yield in the neighborhood.

Location.—While exposure is in some cases an important factor in determining the total yield an acre, and so the cost, the location of the field as regards distance from marketing point and the character of the roads between them is of far greater importance in determining the cost and profit of crop, but one [Pg 39] which is very often disregarded. The marketable product of an acre of tomatoes weighs from 3 to 30 tons, which is not only more than that of most farm crops, but the product is of such character that its value is easily destroyed by long hauls over ordinary roads. It has to be marketed within a day or two of the time it is in prime condition, regardless of the conditions of the roads or weather; so that it is quite deceptive to estimate the cost of delivery at the same rate a ton, as for potatoes or wheat, for it always costs more, and sometimes several times more, to deliver tomatoes than it would to deliver the same weight of less perishable crops. In most cases the cost of picking and delivery is one of the most important factors in determining profit and loss, particularly when the crop is grown for canning factories, where one often has to wait for hours for his team to unload. These conditions make it very important that the field be located within a short distance of, and connected by good roads with the point of delivery.

Early maturing fruit.—Where early maturity is the great desideratum the exposure of the field is often very important. It should, first of all, be such as to secure comparative freedom from spring frosts so as to permit of early setting of the plants and the full benefit of the sunshine as well as protection from cold winds. There is often a great difference in these respects between fields quite near each other. Professor Rolfs, of Florida, mentions a case where the tomatoes in a field sloping to the southeast and protected on the north and west by a strip of oak timber were uninjured by a spring frost that killed not only all [Pg 40] the plants in neighboring fields, but those in the same field farther away from the protecting timber. Such spots should be sought out and utilized, as often they can be used to great advantage. Immediate proximity to large bodies of water is sometimes advantageous in the South, but in the North it is often disadvantageous for early fruit because of the chilling of the air and the increased danger of spring frosts, although affording protection from those of early fall. Here, too, proximity of field to shipping point and distance and transportation rate to market are very important factors affecting profit on the crop.

The home garden.—The south side of buildings or of tight fences and walls often furnishes a most desirable place for garden tomatoes, but the plants should be set at least 6 to 10 feet from the protection and not so as to be trained upon or much shaded by them, as the disadvantage of shutting off the light and circulation of the air, even from the north, would more than overbalance anything gained by the protection.

Growing under glass.—In this country tomatoes are seldom grown under glass except during the darker winter months and the exposure of the house; the form of the roof and the method of glazing which will give the greatest possible light, are of importance, for tomatoes can not be profitably grown in a dark house. Just how the greatest amount of light may be made available in any particular case will depend upon local conditions, but every effort should be made to secure the most unobstructed sunlight possible and for the greatest number of hours each day. [Pg 41]

Previous crop and condition.—In field culture tomatoes should not follow tomatoes or potatoes. Both of these crops make use of large quantities of potash, and although a small part of that used by the plants is taken from the field in the crop, they inevitably reduce the proportion of this element in the soil—that is, in such condition as to be readily available for the succeeding crop. It is true that the deficiency in potash may be supplied, but it is not so easy to supply it in a condition in which it is possible for the roots of the tomato to take it in. Unlike potatoes, tomatoes do not do well on new land, whether it be newly cleared timber lands or new breaking of prairie. Clover leaves the land in better condition for tomatoes than any other of the commonly grown farm crops, while for second choice I prefer one of peas, beans, corn, or wheat in the order named.

One of the most successful tomato growers I know of, whose soil is a rich, dark clay loam, prepares for the crop, as follows: Very late in the fall or early in the spring he gives a clover sod a heavy dressing of manure and plows it under. In the spring he prepares the ground by frequent cultivation and plants it with early sweet corn or summer squash. At the time of the last cultivation of these crops he sows clover seed, covering it with a cultivator having many small teeth, and rarely fails to get a good stand and a good growth of young clover before the ground freezes. In the spring he plows this under, running the plow as deep as possible and following in the furrow with a sub-soiler which stirs, but does not bring the subsoil to the surface. He then gives the field a [Pg 42] heavy dressing with wood ashes and puts it into the best possible tilth before planting his tomatoes. This grown usually harvests at least 500 bushels to the acre and has made a crop of over 1,000 bushels.

Early market.—In some sections of the South where the soil is light and the growers depend almost wholly on the use of large quantities of commercial fertilizer, they seem to meet with the best success by using the same field for several successive crops, but in some places they succeed best with plantings following a crop of cowpeas or other green soiling crops plowed under, with a good dressing of lime.

[Pg 43]

CHAPTER VII

Fertilizers

The experiences and opinions of different gardeners and writers vary greatly as to the amount and kind of fertilizer necessary for the production of the maximum crop of tomatoes. If the question were as to the growth of vine all would agree that the more fertilizer used and the richer the soil, the better. Some growers act as if this were equally true as to fruit, while others declare that one can easily use too much fertilizer and get the ground too rich not only for a maximum but for a profitable crop of fruit. I find that the amount an acre recommended by successful growers varies from 40 tons of well-rotted stable manure, supplemented by 1,000 pounds of complete fertilizer and 1,000 pounds of unleached ashes, to one of only 300 pounds of potato fertilizer.

In my own experience the largest yield that I can recall was produced on what would be called rich land, and the application of fertilizer for the tomato crop was not in excess (unless possibly of potash) of that of the usual annual dressing. I think that in preparing a soil for tomatoes, as in selecting social acquaintances, the "new rich" are to be avoided. A soil which is rich because of judicious manuring and careful cropping for many years can scarcely be too rich, while one that is made rich by a single application of fertilizer, no matter how well proportioned, [Pg 44] may give even a smaller yield of fruit because of its excessive use. Again, the proportions of the various food elements vary greatly in different locations.

Professor Halstead finds that in his section of New Jersey the liberal use of nitrate of soda increases the yield and improves the quality, while in some localities of New York, Ohio, and the West, growers find that the yield of first-class fruit was actually lessened by its use. In some sections of the South liberality in the use of phosphates determines the amount and the quality of the crop, while at other points it seems to be of little value. In my own experience the liberal application of potash, particularly in the form of wood ashes, has more often given good results than the application of any other special fertilizer.

If called upon to name the exact quantity and kind of manure for tomatoes, without any knowledge of the soil or its previous condition, I would say 8 to 10 tons of good stable manure worked into the soil as late as possible in the fall or during the winter and early spring and 300 to 600 pounds of commercial fertilizer, of such composition as to furnish 2 per cent. nitrogen, 6 per cent. phosphoric acid and 8 per cent. potash scattered and worked into the row about the time that the plants are set. The use of a large proportion of nitrogen tends to rank growth of vine and soft, watery fruit. The use of a large proportion of phosphoric acid tends to produce soft fruit with less distinctly acid flavor; of potash, to smaller growth of vine and firm but more acid fruit.

I think that even more than with most crops it will be well for the farmer to experiment to determine the [Pg 45] best and most economical fertilizer for his soil, setting aside five to ten plots of 1 to 4 square rods each and apply nitrate of soda, muriate of potash, wood, ashes, and phosphate alone and in different combinations. The results will suggest the combination which he can use to best advantage. In the majority of cases, however, where the soil is reasonably rich, expenditures for putting the ground in the best possible state of tilth will give larger returns than those for manures in excess of that which the land has usually received in the regular rotation for ordinary farm crops.

For the home garden.—Usually a dressing of wood ashes up to a rate of 1 bushel to the square rod, well worked into the soil before the plants are set, and occasionally watering with liquid manure, will generally give the best returns of any special fertilization, it being assumed that the garden has been well enriched with stable manure.

Tomatoes under glass.—Some growers recommend frequent waterings with liquid manure; others a surface dressing of sheep manure; still others a mulch of moderately well decayed stable manure. Plants growing under glass, particularly in pots or boxes, seem to be benefitted by so heavy a dressing that if applied to plants growing outside it would be likely to give excessive growth of vine with but little fruit.

[Pg 46]

CHAPTER VIII

Preparation of the Soil

The proper preparation of the soil before setting the plants is one of the most essential points in successful tomato culture. The soil should be put into the best possible physical condition and to the greatest practicable depth. How this can be best accomplished will vary greatly with different soils and the facilities at the command of the planter. My practice on a heavy, dry soil is to plow shallow as early in the spring as the ground is fit to work, and then work and re-work the surface so as to make it as fine as possible.

If I am to use any manure which is at all coarse, it is well worked in at this time. A week or 10 days before I expect to set the plants I again plow, and to as great a depth as practicable, without turning up much of the sub-soil, and if this has not been done within two years, follow in the furrows with a sub-soil plow which loosens, but does not bring the sub-soil to the surface. Then I work and re-work the surface, at the same time working in any dressing of well-rotted manure, ashes or commercial fertilizer that I want to use. I never regret going over the field again, if by so doing I can improve its condition in the least. On a lighter soil it might be better to compact rather than loosen as much as would give the best results with clay, but always and everywhere the soil should be [Pg 47] made fine, friable and uniform in condition, to the greatest depth possible.

One of the most successful growers has said that if he could afford to spend but two days' time on a patch of tomatoes he would use a day and a half of the two days in fitting the ground before he set the plants. It is my opinion that any working of the ground that serves to get it into better mechanical condition, if done economically, will not only increase the yield, but to such an extent as to lower the cost a bushel. T. B. Terry's teaching of the necessity for working and re-working the soil, if one would have the largest crops of potatoes of the best quality, is even more applicable to the culture of tomatoes.

Home garden.—Here there is no excuse for setting plants in hard, lumpy soil. It should be worked and re-worked, not simply once or twice, but once or twice after it has been thoroughly worked. In short, the tomato bed should be made as friable as it is possible to make it and to as great a depth as the character of the sub-soil will permit.

Under glass.—I would strongly advise that soil for tomatoes, whether it is to be used in solid beds or in pots or boxes, be thoroughly sterilized by piling it not over 15 inches deep or wide over iron pipes perforated with two lines of holes about one-sixteenth inch in diameter and 2 inches apart and filled with steam for at least a half hour. It can be sterilized, but far less effectively, by thorough wetting with boiling water. It should always be well stirred and aired before the plants are set in it.

Starting plants.—From about the latitude of New [Pg 48] York city southward, it is possible to secure large yields from plants grown from seed sown in place in the field, and one often sees volunteer plants which have sprung up as weeds carrying as much or more fruit than most carefully grown transplanted ones beside them. In many sections tomatoes are grown in large areas for canning factories, and as a farm rather than a market garden crop, individual farmers planting from 10 to 100 acres; and to start and transplant to the field the 25,000 to 30,000 plants necessary for a ten-acre field seems a great undertaking. Tomato plants, however, when young, are of rather weak and tender growth, and need more careful culture than can be readily given in the open field; and, again, the demand of the market, even at the canning factories, is for delivery of the crop earlier than it can be produced by sowing the seed in the field.

For these reasons it is almost the universal custom of successful growers to use plants started under glass or in seed-beds where conditions of heat and moisture can be somewhat under control. I believe, however, that the failure to secure a maximum yield is more often due to defective methods of starting, handling and setting the plants than to any other single cause. In sections where tomatoes are largely grown there are usually men who make a business of starting plants and offering them for sale at prices running from $1 or even as low as 40 cents, up to $8 and $10 a 1,000, according to their age and the way they are grown; but generally, it will be found more advantageous for the planter to start his plants on or near the field where they are to be grown. [Pg 49]

Tomato plants from cuttings may be easily grown, but such plants, when planted in the open ground, do not yield as much fruit as seedlings nor is this apt to be of so good quality; so that, in practice, seedlings only are used for outside crops. Under glass, plants from cuttings do relatively better and some growers prefer them, as they commence to fruit earlier and do not make so rank a growth.

Seedlings can be most easily started and grown, at least up to the time of pricking out, in light, well-ventilated greenhouses, and many large growers have them for this specific purpose. Houses for starting tomato plants should be so situated as to be fully exposed to the sun and not shaded in any way; be provided with heating apparatus by which a night temperature of 60 and up to one of 80° F. in the day can be maintained even in the coldest weather and darkest days likely to occur for 60 to 90 days before the plants can be safely set out in the open field; and the houses should be well glazed and ventilated.

Houses well suited for this purpose are often built of hotbed sash with no frame but a simple ridge-board and sides 1 or 2 feet high, head room being gained by a central sunken path and the sash so fastened in place that they may be easily lifted to give ventilation or entirely removed to give full exposure to sunshine, or for storing when the house is not needed. Hotbed sash 3x6 feet with side-bars projecting at the ends to facilitate fastening them in place are usually kept by dealers, who offer them at from $1.50 to $3 each, according to the quality of the material used.

A hot water heating apparatus is the best, but often [Pg 50] one can use a brick furnace or an iron heating stove, connected with a flue of sewer or drain-pipe that will answer very well and cost much less. It requires but 6 to 10 square feet of bench to start plants enough for an acre, and a house costing only from $25 to $50 will enable one to grow plants enough for 20 acres up to the stage when they can be pricked out into sash or cloth-covered cold-frames in which they can be grown on to the size best suited for setting in the field. When a grower plants less than 5 acres it is often better for him to sow his seed in flats or shallow boxes and arrange to have these cared for in some neighboring greenhouse for the 10 to 20 days before they can be pricked out.

[Pg 51]

CHAPTER IX

Hotbeds and Cold-frames

Plants can be advantageously started and even grown on to the size for setting in open ground in hotbeds. In building these of manure it is important to select a spot where there is no danger of standing water, even after the heaviest rains, and it is well to remove the soil to a depth of 6 inches or 1 foot from a space about 2 feet larger each way than the bed and to build the manure up squarely to a hight of 2 to 3 feet. It is also very important that the bed of manure be of uniform composition as regards mixture of straw and also as to age, density and moisture, so as to secure uniformity in heating. This can be accomplished by shaking out and evenly spreading each forkful and repeatedly and evenly tramping down as the bed is built up. Unless this work is well and carefully done the bed will heat and settle unevenly, making it impossible to secure uniformity of growth in different parts.

Hotbed frames should be of a size to carry four to six 3x6-foot sash, and made of lumber so fastened together that they can be easily knocked apart and stored when not in use. They should be about 10 inches high in front and 16 or 18 inches at the back, care being taken that if the back is made of two boards one of them be narrow and at the bottom so that the crack between them can be covered by banking up [Pg 52] with manure or earth. In placing them on the manure short pieces of board should be laid under the corners to prevent their settling in the manure unevenly. I prefer to sow the seed in flats or shallow boxes filled with rich but sandy and very friable soil, and set these on a layer of sifted coal ashes covering the manure and made perfectly level, but many growers sow on soil resting directly on the manure; if this is done the soil should be light and friable and made perfectly level. A perspective view of a three-sash hotbed is given in Fig. 13, and of a cross-section in Fig. 14.

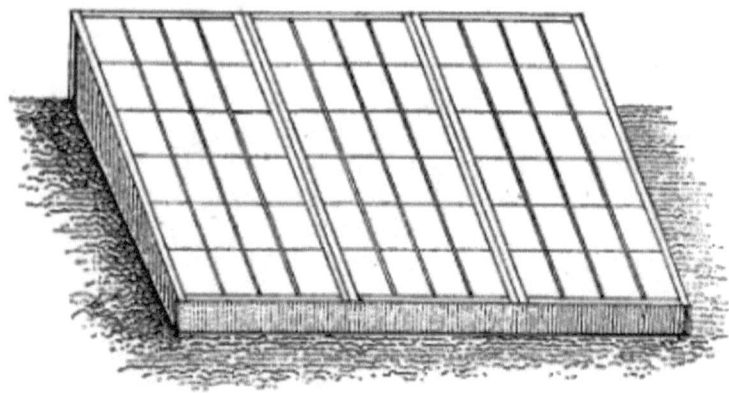

FIG. 13—THREE-SASH HOTBED

In some sections, particularly in the South, it is not always easy to procure suitable manure for making hotbeds, so these are built to be warmed by flues under ground, but I think it much better where a fire is to be used that the sash be built into the form of a house. A hotbed of manure is preferred to a house by some because of its supplying uniform and moist bottom heat—and one can easily give abundant air; [Pg 53] but the sash can be built into the form of a house at but little more expense, and it has the great advantage of enabling one to work among the plants in any weather, while, if properly built, any desired degree of heat and ventilation can be easily secured. Except when very early ripening fruit is the desideratum, plants started with heat but pricked out and grown in coldframes without it, but where they can be protected during cold nights and storms, will give better results than those grown to full size for the field in artificial heat.

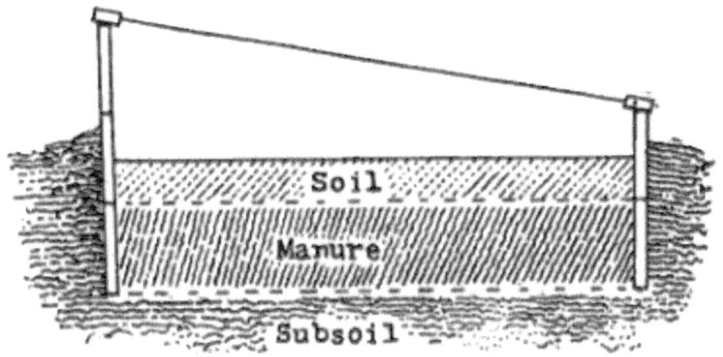

FIG. 14—CROSS-SECTION OF HOTBED

Cold-frames.—In locations where tomatoes are much grown large areas are devoted to cold-frames covered either by sash or cloth curtains. Sash give much better protection from cold and on this account are more desirable, particularly where very early fruiting is wanted, but their first cost is much greater and the labor of attending to beds covered by them is much more than where cloth is used. Sash-covered beds should be of single width and run east and west, but [Pg 54] if the beds are covered with cloth it is better that they be double width (12 feet) and run north and south. The front of the single and the sides of the double width beds should be 8 to 10 inches high, held firmly erect by stakes and perfectly parallel, both horizontally and vertically, with the back or with the central support. This should be 6 inches higher than the front. The cross strips, when sash are used, should be made of a 3-inch horizontal and a 1½-inch vertical strip of 1-inch lumber nailed together very firmly in the form of an inverted T, the vertical pieces projecting 1 inch at each end and resting on the front and back of the bed and forming supports and guides for the sash. Some growers use vertical strips as heavy as 2×3 or 4 inches for stepping across the beds. When [Pg 55] the plants are to be taken to the field, the sash and guides can be easily removed. (Fig. 15.)

FIG. 15 — COLD-FRAMES ON HILL-SIDE

Ground to be covered with cold-frames should be made very friable and rich by repeated plowing and working in of a liberal dressing of well-rotted stable manure and wood ashes. In southwestern New Jersey, where immense areas of early tomatoes are grown, the soil of the beds for a depth of about 6 inches is removed and a layer 3 to 5 inches deep of well-rotted stable manure is placed in. That made of a mixture of manure from horses, cattle and hogs is preferred. It is important that the manure be so well rotted that it will not heat, and so dry that it will not become pasty when tramped into a firm, level layer. On this they place a layer of nearly 3 inches deep of rich, friable, moderately compact soil and prick out the plants into this. The roots soon bind the manure and soil together and by cutting through the manure so as to form blocks one can carry the plants to the fields with but very little disturbance of the root.

Cloth covers for beds should be made of heavy, unbleached sheeting or light duck, and it is better that the selvage run up and down the bed rather than lengthwise. The cloth is torn into lengths of about 13 feet and then sewn together with a narrow double-stitched flat seam so as to form a sheet 13 feet wide and about 8 inches longer than the bed. The edges are tacked every foot to the strips about 2 inches wide by 7/8 inch thick with beveled outside edges and laid perfectly in line. A second line of strips is then nailed to the first so as to break joints with it and so that the two will form a continuous roller about a foot longer [Pg 56] than the bed with the edge of the curtain firmly fastened in its center. The center of the curtain is secured to the central ridge of the bed by strips of lath. When rolled up, the rollers are held in place by loops of rope around their ends and when they are down they are held by similar loops to the notched tent-pins driven into the ground or to wooden buttons fastened to the sides and ends of the frame as shown in Fig. 16.

FIG. 16 — TRANSPLANTING TOMATOES UNDER CLOTH-COVERED FRAMES
(Photo by Prof. W. G. Johnson)

Cloth covers are sometimes dressed with oil, but this is not to be recommended, though it is an advantage to have them wet occasionally with a weak solution of copper sulphate or with sea water as a preser [Pg 57] vative and to prevent mildew. Such covers, well cared for, may last five years or be of little use after the first, depending upon the care given them. They can be made from 50 to 200 feet long and two men can roll them up or down very quickly.

When cloth covers are used the supporting cross-strips should not be over 3 inches wide nor more than 3 feet apart; sometimes the strips are made to bind the sideboard and ridge together by means of short pieces of hoop iron or of barrel hoop. These are so placed and nailed as to hold the upper edge of sideboards and of the central ridge flush with the cross-strips, thus forming a smooth surface for cloth to rest on and enabling one easily to "knock down" and remove the frames to facilitate the taking of the plants from the bed to the field and the storing of the frames for another season.

Flats for starting seeds.—Any shallow box may be used or the plants sown directly in the beds without them, but flats of a uniform size are to be preferred—these will pack well on the greenhouse shelves; or in the hotbed we make them with 7/8 inch thick ends and ½ inch thick sides and bottom, the latter if of a single board having four half-inch holes for drainage and in any case having two narrow strips about ¼ inch thick nailed across their bottoms so as to allow drainage water to escape freely when the boxes are set on hard, cool floors. Two or three such boxes, 35½ inches long, 12 inches wide and 3 inches deep, will be sufficient to start plants enough for an acre. I like to use similar boxes only 4 inches deep for growing the plants after they are pricked out, particularly [Pg 58] if this is to be done in a greenhouse, as by turning them we can equalize exposure to light and thus distribute the plants in the field where they are to be set with the least possible disturbance. One would need nearly 60 such boxes for plants enough for an acre. On account of the lessened necessity for watering when plants are set in beds rather than in boxes, many growers prefer to grow their plants in that way.

[Pg 59]

CHAPTER X

Starting Plants

This has been the subject of a vast amount of horticultural writing, and the practice of different growers, and in different sections, varies greatly. I give the methods I have used successfully, together with reasons for following them, but it may be well for the reader to modify them to suit his own conditions and requirements.

Largest yield.—Some 45 to 50 days before plants can be safely set in the open field the flats in which the seed is to be sown should be filled with light, rich, friable soil, it being important that its surface be made perfectly level, and that it be compact and quite moist, but not so wet as to pack under pressure. Sow the seed in drills 3/8 inch deep and 2 to 3 inches apart at the rate of 10 to 20 to the inch; press the soil evenly over them, water and place in the shade in an even temperature of 80 to 90° F. As soon as the seeds begin to break soil, which they should do in three to four days, place in full light and temperature of 75 to 80°, keeping the air rather close so as to avoid necessity of watering. After a few days reduce the temperature to about 65° and give as much air as possible. Some growers press a short piece of 2-inch joist into the soil of the benches, so as to form trenches 2 inches wide and about 3/8 inch deep, and so spaced as to be under the center of each row of glass, their sash being [Pg 60] mostly made of five-inch glass. In this, by using a little tin box with holes in the top, like those of a pepper-box, they scatter seeds so that they will be nearly 1/8 to ¼ inch apart, over the bottom of the 2-inch wide trench, and then cover. This has the advantage of evenly spacing the plants and so locating the rows that the plants will be little liable to injury from drip.

Young tomato plants are very sensitive to over-supply of water and some of the most successful growers do not water at all until the plants are quite large and then only when necessary to prevent wilting. In 10 to 15 days, or as soon as the central bud is well started, the plants should be pricked out, setting them 3 to 6 inches apart, according to the size we expect them to reach before they go into the field; 5 inches is the most common distance used. I think it better to set the full distance apart at first, not to transplant a second time. It is very important that this pricking out should be done when the plants are young and small, though many successful growers wait until they are larger. The soil in which they are set, whether it be in boxes or beds, should be composed of about three parts garden loam, two parts well-rotted stable manure and one part of an equal mixture of sand and leaf mold, though the proportion of sand used should be increased if the garden loam is clayey. The soil in the seed-boxes or in the beds, when the seedlings are taken up, should be in such condition, and the plants be handled in such a way that nearly all the roots, carrying with them many particles of soil, are saved. The plants should be set a little, and but a little, deeper than they stood in the seed-box and the soil so pressed about the [Pg 61] roots, particularly at their lower end, that the plants can not be easily pulled out.

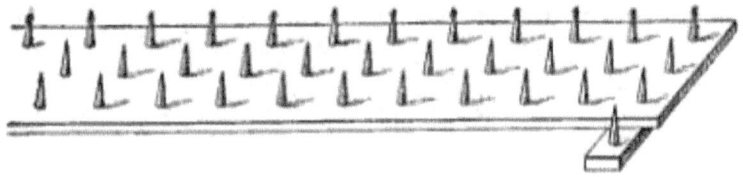

FIG. 17—SPOTTING-BOARD FOR USE IN COLD-FRAMES

Where plants are set in beds the work can be facilitated by the use of a "spotting-board" (Fig. 17). This should be about 1 foot in width, and have pegs about 3 inches long, ¾ inch in diameter at the base and tapering to a point, fastened into the board the distance apart the plants are to be set. It should also have narrow projections carrying a single peg nailed to the top of board at each end, so that when these pegs are placed in the end holes of the last row the first row of pegs in the "spotting board" will be the right distance from the last row of holes or plants. By standing on this, while setting plants in one set of holes, holes for another set are formed. If the conditions of soil, air and plants are right and the work is well done, the plants will show little tendency to wilt, and it is better to prevent their doing so by shading, rather than by watering, though the latter should be resorted to if necessary. When plants are set in beds, some growers remove the soil to a depth of about 6 inches and put in a layer of about 2 inches of sifted coal ashes, made perfectly level, and then replace the soil. This confines the roots to the surface and enables one to secure nearly all of them when transplanting. The [Pg 62] plants should be well established in 24 hours and after this the more light and air that can be given, without the temperature falling below 40° F. or subjecting the plants to cold, dry wind, the better.

FIG. 18—SPOTTING-BOARD FOR USE ON FLAT
(From W. G. Johnson)

One can hardly overstate the importance to the healthy growth of the young tomato plant of abundant sunshine, a uniform day temperature of from 60 to 80° F., or of the ill effects of a variable temperature, particularly if it be the result of cold, dry winds, or of a wet, soggy soil, the effect of over-watering. These points should be kept in mind in caring for the plants, and every effort made to secure, as far as possible, the first named conditions and to avoid the latter. The frames, whether they be covered with sash or cloth, but more particularly if with sash in sunshine and with curtains in dull days, should be opened so as to prevent their becoming too hot, and so as to admit air. And in a greenhouse full ventilation should [Pg 63] be given whenever it is possible to do so without exposure to too low a temperature. If the plants are in boxes and on greenhouse shelves, it is important that these be turned end for end every few days to equalize exposure to light and give full exposure to the sun. The plants should be watered only when necessary to prevent wilting, and the beds should be covered during heavy rains. A "spotting-board" for use on flats is seen in Fig. 18.

The most unfavorable weather conditions are bright sun combined with a cold wind, and cold storms of drizzling rain and frosty nights. Loss from the latter cause may often be prevented by covering the beds with coarse straw, which should always be provided for use in an emergency. Many growers provide a second curtain—an old one answers very well—to throw over the straw-covered beds. Beds so covered will protect the plants from frost in quite severe weather. Watering should especially be avoided for nearly three days before setting in fields; but six to twelve hours before it is well to water thoroughly, though not so as to make the soil at all muddy. About five days after pricking out and again about five days before the plants are to go into the field and five days after they are set, they should be sprayed with Bordeaux mixture.

Early ripening fruit.—Here the aim is to secure, by the time they can be set in the field, plants which have come by an unchecked but comparatively slow rate of growth to the greatest size and maturity consistent with the transplanting to the field without too serious a check. The methods by which this is accomplished [Pg 64] vary greatly and generally differ materially from those given above. The seed is planted much earlier and 60 to 90 days before it is at all safe to set plants in the open field; while a steady rate of growth is desirable, it should be slow and the plants kept small by a second and even third and fourth transplanting, and especial care taken to avoid the soft and irregular growth resulting from over-watering or over-heating. Any side shoots which may appear should be pinched out and a full pollination of the first cluster of the blossoms secured, either by direct application of pollen or by staking or jarring the plants on bright days; and finally, special efforts made to set the plants in the field as early and with as little check as possible. Growers are often willing to run considerable risk of frost for the sake of early setting.

When one has sandy land a very profitable crop can sometimes be secured by sowing the seed very early, and growing the plants on in beds until the first cluster of fruit is set, then heeling them in, much as nursery trees are, but so close that they can be quickly covered in case of frost. As soon as it is at all safe to do so, they are set in the open ground, very closely, on the south side of ridges, so that only the upper one-third of the plant is exposed, the remainder being laid nearly level and covered with earth.

So treated the plants will ripen the upper one or two clusters very early but will yield little more until late in the season, and it is generally more profitable to plow them up and put in some other crop as soon as the first clusters of fruit have ripened. Others pinch out the central bud as soon as it is well formed, [Pg 65] usually within 10 days from the sowing of the seed. When this is done a great proportion of the plants will start branches from the axils of the cotyledons; these usually develop blossoms in the third to the fifth node and produce fruit much lower than in a normal plant. It is questionable if there is any gain in time from seed to fruit by this method, but it enables one to get older plants of a size which it is practicable to transplant to the field.

In most cases it will be found more profitable and satisfactory so to grow the plants that by the time they can be safely set out of doors they will be in vigorous condition, about 6 to 10 inches tall, stout, healthy and well hardened off. Such plants will ripen fruit nearly, and often quite as early as older ones and will produce a constant succession of fruit, instead of ripening a single cluster or two and then no more until they have made a new growth.

For late summer and early fall.—It is generally true in the South and often equally so in the North, that there is a more eager local demand for tomatoes in the late summer and fall months, after most of the spring set plants have ceased bearing, than in early summer. In Michigan I have often been able to get more for choice fruit in late October and in November than the best Floridas were sold for in May or early June, and certainly in the South the home use of fresh tomatoes should not be confined to spring set plants. For the fall crop in the South seed may be sown in late spring or up to the middle of July, in beds shaded with frames, covered with lath nailed 3 to 4 inches apart and the plants set in the field about 40 days from [Pg 66] sowing, the same care being taken to put the ground into good condition as is recommended for the spring planted crop.

A second plan, which has sometimes given most excellent results, is to cut back spring set plants which have ripened some fruit but which are not completely exhausted, to mere stubs, and spade up the ground about them so as to cut most of the roots, water thoroughly and cover the ground with a mulch of straw. Most of the plants so treated will start a new and vigorous growth and give most satisfactory returns.

Fruit at least expenditure of labor.—When this is the great desideratum, many growers omit the hotbed and even the pricking out, sowing the seed as early as they judge the plants will be safe from frost, and broadcast, either in cold-frames or in uncovered beds, at the rate of 50 to 150 to the square foot and transplanting directly to the field. Or they may be advantageously sown in broad drills either by the use of the pepper-box arrangement suggested on page 60, or a garden drill adjusted to sow a broad row. In Maryland and the adjoining states, as well as in some places in the West, most of the plants for crops for the canners are grown in this way and at a cost of 40 cents or even less a 1,000. The seed should be sown so that it will be from ¼ to ½ inch apart and the plants thinned as soon as they are up so that they will be at least ½ inch apart. Where seed is sown early with no provision for protection from the frost it is always well to make other sowings as soon as the last begins to break ground in order to furnish reserve plants, if the earlier sown lots be destroyed by frost. Others [Pg 67] even sow the seed in place in the field, thinning out to a single one in a hill when the plants are about 2 inches high. Some of the largest yields I have ever known have been raised in this way, but the fruit is late in maturing and generally the method is not so satisfactory as starting the plants where they can be given some protection, and transplanting them to the field.

Plants for the home garden.—These may be grown in pots or boxes set in the sunniest spot available and treated as has been described. In this way plants, equal to any, may be grown without the aid of either hotbed or greenhouse. It will generally be more satisfactory, however, to secure the dozen or two plants needed from some one who has grown them in quantity than to grow so small a lot by themselves. In selecting plants, take those which are short, stiff, hard, and dark green in color with some purple color on the lower part of the stem rather than those which are softer and of a brighter green, or those in which the foliage is of a yellowish green; but in selection it must be remembered that varieties differ as to the color of foliage, so that there may be a difference in shade which is not due to conditions.

Plants under glass.—If to be grown in pots or boxes, "prick out," when small, into three-inch pots and as they grow re-pot several times so that when set in the pots or beds in which they are to fruit, they are stout plants 12 to 16 inches high. Plants propagated from cuttings give much better returns relatively under glass than out of doors.

[Pg 68]

CHAPTER XI

Proper Distance for Planting

The best distance apart for the plants to be set in the field varies greatly with the soil, the variety, the methods of cultivation and other conditions. Plants set as close in rich clay soil as would give the best results in a warm, sandy one, or those of a strong growing sort, like Buckeye State, set as close as would be desirable for sorts, like Atlantic Prize or Dwarf Champion, would give little but leaves and inferior fruit. In field culture I like to space the plants so as to facilitate gathering the fruit, and recommend the following arrangement: Set the plants according to soil and the variety 2½ to 4 feet apart in the row, omitting two or three in every 75 or 100 plants so as to form driveways across the rows. Set the first and second and the third and fourth rows, etc., 2½ to 3½ and the second and third and the fourth and fifth rows 5½ to 6 feet apart. As the plants grow, those of the first and second and those of the third and fourth rows, etc., are thrown together and in many cases it will pay to have a pair of narrow horizontal strips or wires nearly 18 inches from the ground upon which they can be thrown.

This arrangement of the plants allows us to continue to cultivate the wider spaces between the second and third and fourth and fifth, etc., rows, much longer, and tends to confine the necessary tramping and pack [Pg 69] ing of the soil when gathering the fruit chiefly to these rows—an important point in case the soil is wet. The rows can be marked out the day before, but it is better to set the plants in the cross-rows and that these be marked out just ahead of the setters. In this arrangement the distances are equivalent to from 2½×4 feet, requiring 4,300 plants to the acre, to 4×5 feet, requiring but about 2,100 plants. The latter distance is that most commonly used by New Jersey growers.

FIG. 19—TOMATOES SOWN AND ALLOWED TO GROW IN HOTBEDS

In the home garden.—It will usually be more satisfactory to give each plant plenty of space, setting them 5 or 6 feet apart each way, except in the case of the dwarf sorts, which should be from 3½ to 3 feet apart. A few plants at these distances will usually be much more satisfactory than more set nearer together, but [Pg 70] the larger growing sorts should have at least 3 feet and the dwarf sorts 2 feet. When one has a hotbed or cold-frame it is often an advantage to set a row of tomato plants nearly 18 inches apart at the back end much earlier than they could be safely set in the open ground, and if these are allowed to grow on in place, as shown in Fig. 19, being pruned and tied to stakes, they will give some very early fruit.

In the greenhouse.—Experience and practice differ as to the most desirable distance apart for plants under glass. But 2 feet apart, where quality is the main consideration, and 18 inches when quantity, if fair, is of more importance than extra quality.

Setting plants in the field.—The economical and successful setting of plants in the field is an important element of successful tomato culture and is very dependent upon soil and weather conditions. It is assumed that the soil of the field has been put into the best possible condition of tilth, but its condition as to moisture is also very important. The worst condition is when it is wet and muddy, especially if it is at all clayey—not only is the cost of setting greatly increased, but plants set in such soil can seldom, by any amount of care, be made to do well, especially if a heavy beating rain or dry windy weather follows immediately; the condition is less unfavorable if a warm gentle rain or still moist weather follows. A dry cold wind, even if the day is cloudy and the soil in good condition, is also unfavorable, particularly if the roots of the plants are exposed.

Wet soil, cold, dry air and wind are the conditions to be avoided. Moist, not wet, soil and still, warm air [Pg 71] are to be desired; whether the day is sunny or not is less important. There is a certain definite time, which does not usually extend beyond a few days, when any lot of plants is in the best condition for setting in the field. It is hardly possible to describe this condition more than to say it is when the plants are as large as they can be without crowding and are in a state where they can best stand the shock of removal.

It will always be a matter of judgment as to how long it is best to hold plants, which are in condition for setting, for favorable weather conditions. They can sometimes be held a few days, by scant watering and full exposure, or in some cases by taking from the bed and heeling in, as nurserymen do trees; but it is better to set when the weather is unfavorable or to run some risk from frost rather than to hold them in this way too long. The wise selection of time for setting is an important factor in securing a good and profitable crop.

The South Jersey growers, to whom early ripening fruit is the great desideratum and who have a very warm soil, and grow plants so they are quite hardy and can be transplanted with little check, set them in the field very early, some seasons by the last of April; and if the plants can be got out so as to have two or three days of favorable weather to get established before it comes, they seem to be little hurt even by a quite severe frost. The first essential to successful transplanting is to have well-grown, healthy, hardy plants; the second is that they be in good condition for setting, which can be secured by giving them, for a few days before planting, a scant supply of water and [Pg 72] fullest possible exposure to air and sun, and then a thorough wetting a few hours before they are to be set.

The South Jersey plan of growing and setting plants gets them into the field in the best condition of any method I know. Two to five days before they expect to plant, the growers go over the beds and, by means of a hoe that has been straightened and sharpened to form a sort of spade, they cut through the soil and manure so as to divide the plants into blocks of six. A few hours before they are to plant, they saturate the bed with water. By means of a flattened shovel they can take up the blocks of plants and place them in a cart or low wagon so the soil is scarcely disturbed at all, the roots in the manure serving to bind the whole together. In the meantime furrows are opened along the rows and the cart driven to the field; the plants in the blocks are cut apart with a butcher knife placed in the furrow and the earth drawn about them.

Plants set in this way often do not wilt at all, even in hot sunshine. When plants are grown in boxes these can be taken to the field and plants taken from them in much the same way and so that they will be disturbed but little. In setting the plants it should always be borne in mind that while sunshine on the leaves of a plant rarely does any injury, it is very injurious to the roots, and the exposure of the roots to the sun or to cold, dry wind, should be avoided in every practicable way, such as by carrying the plants to the field laid on the sides of a box, which is then carried with its bottom toward the sun so as to have the plants in the shade, always handling the plant in the shade of one's body, etc. It is well worth while [Pg 73] to walk to the end of the row to commence work in order to secure this. It is attention to such details that distinguishes one whose plants nearly always do well from one who loses a large proportion of those he handles.

Fruit at the least expenditure of labor. — The plants are prepared for setting by scant watering, and are taken up so as to secure as much root as possible with little soil adhering to them. Great care should be taken in taking the plants from the bed, and in handling them, *to avoid twisting the stems*, as to do so very seriously injures the plants, often to such an extent that they will fail to grow, no matter how carefully set out. Some growers dip the roots in a very thin clay mud, hardly thicker than thin cream, but I have not found this of advantage except, sometimes, when the roots are to be exposed for a longer period than usual and I do not recommend it for general use. In setting, holes are made either with a long dibble, in the hands of the one who distributes the plants, or by a short one, in the hands of the setter; the plants are dropped into them a little deeper than they had stood in the bed, the earth closed about the roots, by pressure from the side. Especial care should be taken that this is well done, particularly at the bottom; the earth should be so firmly pressed to the root that the plant cannot be easily pulled from the soil. In some sections transplanting machines (Fig. 20) are used and liked, but most planters prefer to set by hand and the additional cost is not great. An expert with one or two boys to assist in handling the plants can put out as many as 5,000 plants in a day. A machine re [Pg 74] quiring more help to run it can set from 15,000 to 20,000.

In the home garden, when but a few plants are to be set, it will be better to put them in after 4 P. M. and use water in setting, but the wet soil should be covered with some dry earth to prevent its caking.

In the greenhouse.—Plants are better set in the places where they are to fruit just before their first blossoms open and should be set in accordance with the suggestions given for transplanting to the field.

FIG. 20 – PLANTING TOMATOES ON A DELAWARE FARM
(Photo by courtesy of *American Agriculturist*)

CHAPTER XII

Cultivation

For maximum crop. — As soon as plants are set the ground should be well cultivated to the greatest depth practicable. We should remember that the tomato needs for its best development a very friable soil, while the tramping necessary in setting out the plants and gathering the fruit tends to compact and harden the soil. Often transplanting has to be done when the soil is wet, and we need to counteract the injury from tramping by immediate cultivation; but, at the same time, we must avoid the disturbing of the plants any more than is necessary, and all of our cultivation should be done with these points in mind. Just how it can be done best will vary not only with the location and the facilities available, but with the weather conditions, so that it is not well to attempt to give explicit directions any further than that one can hardly cultivate too deeply for the first seven days nor too often for the first 30 days after the plants are set, provided he avoids turning the soil when it is too wet. Even walking through the field when the soil is wet is injurious and should be avoided, in proportion as the soil is a clayey one.

At least expenditure of labor. — I hardly need add to or change the suggestions given above for tomatoes at least cost, for any cultivation wisely given will probably do as much to reduce cost per bushel by [Pg 77] increasing the yield per acre as any other expenditure. *In the garden* it is advisable that from the time the plants are set until the fruit ripens, the surface soil about them be stirred every evening when it is not actually wet.

In the greenhouse. — The surface of the soil should be kept open by frequent stirring or, as is the practice of some successful growers, it may be covered with a mulch of partially rotted manure. The plants should be watered only as needed to prevent wilt, and special pains taken to guard against too much moisture either in the soil or in the air, particularly on dark days. The night temperature should be uniformly about 60° F. while in the day it should be 75°, and if it be bright and sunny it may go to 90° or even higher. Air should be given freely whenever feasible to do so without too greatly reducing temperature. A moderate degree of moisture should be maintained in the air, care being taken that it does not become too moist, especially during dark days. There is more danger from the air becoming too moist than from its becoming too dry, though either extreme is injurious.

Pollinating.—The structure and relations of the parts of the tomato flower are such that while perfect pollination is possible, and in plants grown in the open air usually takes place without artificial assistance, it is not so likely to occur when plants are grown under glass, particularly in the winter months, and it is usually necessary to secure it by artificial means. With vigorous, healthy plants and on light, sunny days, it can be accomplished by jarring the plants near midday. This generally throws enough pollen into the [Pg 78] air so that an abundance of it reaches each receptive stigma. With less vigorous plants and on dark days it is necessary to hand pollinate the flowers. This is done by gathering the pollen by means of jarring the plants, so that it falls into a watch crystal or other receptacle secured at the end of a wand, and then pressing the projecting pistils of other flowers into it so that they may become covered with the pollen.

Some growers transfer the pollen with a camel's-hair-brush; others by pulling off the corolla and adhering anthers and rubbing them over the stigma of other flowers. Fruit rarely follows flowers that are not pollinated, and if it is incomplete the fruit will be unsymmetrical and imperfectly developed. As tomato flowers secrete but very little, if any, honey and are not attractive to insects, it is of no advantage to confine a hive of bees in the tomato house in the way which is so useful in one where cucumbers or melons are growing.

[Pg 79]

CHAPTER XIII

Staking, Training and Pruning

Under favorable conditions of soil and climate, plants of most varieties of tomatoes will, in field culture, yield as much fruit if allowed to grow naturally and unpruned as if trained and pruned. This is especially true of the sorts of the Earliana type and on warm, sandy soils, while it may not be true of the stronger growing sorts, or on rich clay lands or where the fertilizer used contains an excess of nitrogen. In any case more fruit can be grown to the acre on pruned and staked plants because more of them can be gotten on an acre; and it is an advantage to grow them in that way because it enables us, by later cultivation, to keep the ground in good tilth longer; also it facilitates the gathering of the fruit; and last, but not least, it generally enables us to produce better ripened and flavored fruit.

Staking and pruning used to be the almost universal practice in the South, but in many sections growers have abandoned it, claiming that they get as good or better results without it. In the North it is rarely used in field culture, though often used in private gardens and by some market gardeners, and both staking or tying up and pruning are essential to the profitable growing of tomatoes under glass. In the South, stout stakes from 1 to 2 inches in diameter and 4 to 5 feet long are driven into the ground so that they can be [Pg 80] depended upon to hold the plants erect through the heaviest storms, as seen in Fig. 21. This is generally and wisely done as soon as the plant is set, though some growers delay doing so until the fruit is well set, claiming that the disturbance of staking, tying and pruning tends to hasten the ripening of the fruit. The plant is then tied up, the tying material being wrapped once about the stake and then looped about the plant so as to prevent slipping on the stake or choking the stem of the plant as it enlarges. Raffia is largely used and is one of the best tying materials, but short pieces of any soft, cheap string can be used. The tying up will need to be repeated as the stem elongates, which it will do very rapidly.

In pruning the tomato we should allow the central shoot of the young plant to grow, and remove all of the side shoots which spring from the axils of the leaves and sometimes even from the fruit clusters, as seen in Fig. 22. It is very desirable that this be done when the branches are small, as there is then less danger of seriously disturbing the balance of the growing forces of the plant, and also because there is less danger of careless workmen cutting off the main shoot in place of a lateral, which would seriously check the ripening of the fruit. It is especially important that any shoots springing from the fruit cluster be removed as early as possible. For these reasons it is important that, if the plants are to be pruned at all, the field be gone over every few days. If the pruning is not well done it is a disadvantage rather than a help.

[Pg 81]

FIG. 21 – TRAINING TOMATOES IN FLORIDA TO SINGLE STAKE
(Photo by courtesy of Prof. P. H. Rolfs, Director Florida Experiment Station)

FIG. 22 – TOMATO PLANT TRAINED TO SINGLE STAKE

FIG. 23—METHOD OF TRAINING TO THREE STEMS IN FORCING-HOUSE AND OUT OF DOORS

[Pg 84] Some growers allow two or three (Fig. 23) instead of one shoot to grow, selecting for the second the most vigorous of the shoots starting from below the first cluster of fruit. In some locations they stop or pinch out the main shoot just above the first leaf above the third or fourth cluster; in some soils it is an advantage and in others rather a disadvantage to do this. I have seldom practiced it. When fruit at the lowest cost a bushel is the desideratum, neither pruning nor staking is desirable.

FIG. 24 — METHOD OF TRAINING ON LINE IN GREENHOUSE

FIG. 25 — READY TO TRANSPLANT IN GREENHOUSE
(Redrawn from photo by New York Experiment Station)

FIG. 26 — TRAINING YOUNG TOMATOES IN GREENHOUSE AT NEW YORK EXPERIMENT STATION
(Photo by courtesy Prof. U. P. Hedrick)

For home gardens. — In the home garden trellising and pruning are often very desirable, as they enable us not only to produce more fruit in a given area but of better quality. Many forms of trellis, have been recommended. Where the plants are to be pruned as well as supported, as they should always be in gardens, there is nothing better than the single stake, as described above. For a trellis without pruning, one to [Pg 87] three stout hoops supported by three stakes so as to surround the plant which is allowed to grow through and fall over them, or two or more parallel strips supported about a foot from the ground on each side of a row of plants answer the purpose, which is simply to keep the plant up from the ground and facilitate the free circulation of the air among leaves and fruit.

FIG. 27—TOMATOES IN GREENHOUSE AT OHIO EXPERIMENT STATION
(Photo by courtesy of C. W. Waid)

I have seen tomatoes grown very successfully by the side of an open fence. Two stakes were driven into the ground about 6 inches from the fence and the plant, but slanting outward and away from each other. The tops of the stakes were fastened to the fence by wooden braces, and then heavy strings fastened to the fence around the stakes and back to the [Pg 88] fence, the whole with the fence forming a sort of inverted pyramidal vase about 3 feet across at the top. In this the plant was allowed to grow, but it would be essential to success that the fence be an open one.

FIG. 28 – FORCING TOMATOES IN GREENHOUSE AT NEW HAMPSHIRE EXPERIMENT STATION. NOTE CHARACTER OF BED ON THE GROUND FLOOR.
(Photo by courtesy of Prof. H. F. Hall)

In the greenhouse. — Here pruning and training are essential. The plants may be supported by wires or strings (a coarse wool twine will answer), twisting the string about the plant as it grows. The growth is usually confined to a single shoot, though some growers allow two (Fig. 24); the method of pruning does not differ from that given for field cul [Pg 89] ture, but it is more important that the plants be gone over often and the branches removed when small. If allowed to do so, branches would spring from the axil of each leaf and the plant would become a perfect thicket of slender branches and leaves and produce but little fruit. The main stem is sometimes pinched out after three or four clusters of fruit are set and the branch from the axil of the first leaf above is allowed to take its place. This tends to hasten the maturing of the fruit clusters already set. After several clusters have matured, or the main stem reaches the top of the house, some growers allow a shoot from the bottom to grow and as soon as fruit sets on it the first stem is cut away and this takes its place. Others prefer to remove the old plant entirely and set in young ones. A plant ready for transplanting is shown in Fig. 25. In figures 26, 27 and 28 are shown interior views of greenhouses at the New York station at Geneva, the Ohio station at Wooster, and the New Hampshire station at Durham. Note the strong, vigorous plants in Fig. 26; the method of utilizing tile for watering in Fig. 27; and the ground-floor bedding in Fig. 28.

[Pg 90]

CHAPTER XIV

Ripening, Gathering, Handling and Marketing the Fruit

Tomatoes ripen and color from within outward and they will acquire full and often superior color, particularly about the stems, if, as soon as they have acquired full size and the ripening process has fairly commenced, they are picked and spread out in the sunshine. The point of ripeness when they can be safely picked is indicated by the surface color changing from a dark green to one of distinctly lighter shade with a very light tinge of pink. Fruit picked in this stage of maturity may be wrapped in paper and shipped 1,000 or 2,000 miles and when unwrapped after two or ten days' journey will be found to have acquired a beautiful color, often even more brilliant than that of a companion fruit left on the vine. Enclosing the fruit while on the vine and about half grown in paper bags has been recommended, and it often results in deeper and more even coloring and prevents injury from cracking, but the fruit so ripened, while more beautiful, is not so well flavored as that ripened in the sun. But Americans are said to taste with their eyes, so that in this country, fruit of this beautiful color will often out-sell that which is of better flavor though of duller color.

The tomato never acquires its full and most perfect flavor except when ripened on the vine and in full [Pg 91] sunlight. Vine and sun-ripened tomatoes, like tree-ripened peaches, are vastly better flavored than those artificially ripened. This is the chief reason why tomatoes grown in hothouses in the vicinity are so much superior to those shipped in from farther south. After it has come to its most perfect condition on the plant the fruit deteriorates steadily, whether gathered or allowed to remain on the vine, and the more rapidly in proportion as the air is hot and moist. That it be fresh is hardly less essential to the first quality in a tomato than it is to such things as lettuce and cucumbers.

Gathering.—As is the case with most horticultural products, the best methods of gathering, handling and marketing the fruit vary greatly with the conditions under which the fruit was grown and how it is to be used, and it requires the best of judgment to gather it in the stage of maturity in which it will give the best satisfaction, under the conditions and for the purposes for which it is to be used. It is impossible to give exact rules for determining when the fruit is in the best condition. This can only be learned by experience, guided by a knowledge of the ripening habit of the fruit, which not only varies somewhat in different localities, but with different varieties. In the extreme South, fruit is picked for shipment before it shows more than the slightest tint of color at the blossom end; the depth of color which is considered as indicating shipping condition deepens as we go north and nearer market.

Generally the fruit should be left on the vine no longer than will permit of its becoming fully ripe [Pg 92] by the time it reaches its destination and is exposed for sale. When the fruit is to be shipped any distance the field should be gone over frequently, as often as every second or third day or even every day in the hight of the season, and care taken to pick every fruit as soon as it is in proper condition. When it is to be sold in nearby markets or to a cannery the exact stage of maturity, when picked, is not so important, although it is always an advantage not to gather until the fruit is well colored and before it begins to soften. Some growers for canneries make but three or four pickings, but in this case it is well to gather the ripest fruit separately.

In picking and handling great care should be taken not to mar or bruise the fruit, and the stems should be removed as the fruit is picked to prevent bruising in handling. A bruise or mar may not be as conspicuous in a tomato as in a peach, but it is quite as injurious. It is a great deal better for pickers to use light pails rather than baskets, the flexibility of the latter often resulting in bruises. It is an advantage to have enough of these so that the sorting can be from the pail, but if this is not practical the fruit should be carefully emptied on a sorting table for grading. It should first of all be separated with regard to its maturity. A single fruit which is a little riper or greener than the remainder may make the entire package unsalable. It should also be graded as to freedom from blemishes or cracks, and as to size, form and color. It is assumed that the fruit for each package is to be of the same variety, but often there is quite a variation in different fruits from even the [Pg 93] same vine; the more uniform in all respects the fruit in a package is the more attractive and salable it becomes. There is no fruit where careful grading and packing have more influence on the price it will command.

FIG. 29 — FLORIDA TOMATOES PROPERLY WRAPPED FOR LONG SHIPMENT
(Photo by courtesy of *American Agriculturist*)

I know of a certain noted peach-grower in northern Michigan who grew, each year, some 2 to 5 acres of tomatoes for the Chicago market. It was his custom to pick out about one-tenth of the best of the fruit, putting it into small and attractively labeled [Pg 94] packages; the remainder of the crop was sorted over and from one-tenth to one-fifth of it rejected and fed to stock or sold to a local cannery. The remainder was sent to Chicago with his selects, but as common stock, and usually brought more than his neighbors received for unsorted fruit; but the check he received for his selects was usually as large as that for his commons, thus giving him about 33-1/3 per cent. more for his crop than his neighbors received for their equally good, but unsorted, fruit—to say nothing of what he received for the rejected fruit and the saving of freight which, he said, was usually enough to pay the actual cost of sorting.

Tomatoes are usually classed as vegetables but, when ripe, they require as careful handling as the most delicate fruits and are as easily and seriously injured by bruising and jarring. Just how this can be avoided and the fruit gotten from the vine to the possibly distant consumer in the best condition will vary in different cases. Tomatoes from the South (Fig. 29) are generally marketed in carriers which, though varying somewhat, are essentially alike and consist of an open basket or boxes of veneer holding about 10 pounds of fruit. When shipped, two, four or six of these are packed in crates made of thin boards, so as to protect the fruits but give them plenty of air.

Packing.—Most of the fruit sent to New York and Philadelphia markets from New Jersey and other northern states is in boxes or crates holding about 5/8 of a bushel and so made as to facilitate ventilation when piled in cars or warehouses. Fruit for the [Pg 95] canneries is usually picked and handled in bushel crates of lath. These various packages are usually sold in the flat and the grower puts them together as is convenient before the crop comes on; but in many sections where there are large shipments they are often put together by the package dealers. Fig. 30 shows tomatoes as packed by the Ohio experiment station.

FIG. 30—GREENHOUSE TOMATOES PACKED FOR MARKET
(By courtesy Ohio Experiment Station)

Fruits after frost.—Sometimes when there is a great quantity of partially ripe and full grown green fruit on the vines which is liable to be spoiled by an early fall frost, it can be saved by pulling the vines and placing them in windrows and covering them with straw. Of course the vines should be handled carefully to shake off as little fruit as possible. If the freeze is followed by a spell of warm, dry weather [Pg 96] the fruit will ripen up so as to be quite equal to that shipped in from a distance. A second plan is to pull the vines and hang them up in a dry cellar or out-house, or lay them on the ground in an open grove of trees, or beneath the trees of an adjoining orchard.

Still another plan is to gather the green fruit and spread it not more than two to four fruits deep in hotbed frames, which are then covered with sash. Local grocers are usually glad to pay good prices for this late fruit, and in seasons of scarcity I have known canners to buy thousands of bushels so ripened at better prices than they paid for the main crop.

[Pg 97]

CHAPTER XV

Adaptation of Varieties

Whatever may be their botanical origin, the modern varieties of cultivated tomatoes vary greatly in many respects, and while these differences are always of importance their relative importance differs with conditions. When the great desideratum is the largest possible yield of salable fruit at the least expenditure of labor, the qualities of the vine may be the most important ones to be considered, while in private gardens and for a critical home market and where closer attention and better cultivation can be given, they may be of far less importance than qualities of fruit.

Habits of growth.—Whether it be standard or dwarf, compact or spreading, is sometimes of great importance as fitting the sorts for certain soils and methods of culture. On heavy, moist, rich land, where staking and pruning are essential to the production of fruit of the best quality, it is of importance that we use sorts whose habits of growth fit them for it; while on warm, sandy, well-drained land, staking and pruning may be of little value, and a different habit of growth more desirable. We have sorts in which the vine is relatively strong growing with few branches, upright, with long nodes and small fruit clusters well scattered over the vine. They are usually very productive through a long season but generally late in maturing. Stocks of this type are sometimes sold, [Pg 98] I think improperly, as giant climbing, or Tree tomato. The Buckeye State is a good type of these sorts. (Fig. 31.)

FIG. 31—BUCKEYE STATE, SHOWING LONG NODES AND DISTANCE BETWEEN FRUIT CLUSTERS

Other varieties make a stout and vigorous but shorter growth, with more and heavier branches, shorter nodes and many small medium-sized clusters of fruit well distributed over the plant and which mature through a fairly long season. These sorts are usually very productive and our most popular varieties generally belong to this type, of which the Stone (Fig. 32) is a good representative of the more compact and the Beauty of the more open growing.

FIG. 32 — STONE, AND CHARACTERISTIC FOLIAGE

Other varieties form many short, weak, sprawling [Pg 100] branches, with usually large and sometimes very large clusters of fruit produced chiefly near the center of the plant and which mature early and all together. Plants of this type will often mature their entire crop and die by the time those of the first type have come into full crop. The Atlantic Prize (Fig. 33) and Sparks Earliana are examples of this type.

In sharp contrast with the above is the tomato De Laye, often called Tree tomato. This originated about 1862 in a garden at Chateau de Laye, France. In this the plant rarely exceeds 18 inches in hight, is single-stemmed or with few very short branches, the nodes very short, the fruit clusters few and small. From this, by crossing with other types, there has been developed a distinct class of dwarf tomatoes which are of intermediate form and character and are well represented by the Dwarf Champion (Fig. 34). Early maturity is sometimes the most important consideration of all, though, because of increasing facilities for shipping from the South, it is less commonly so than formerly. For shipping and canning it is generally, though not always, desirable that the crop mature as nearly together as possible, that it may be gathered with the fewest number of pickings and advantage taken of a favorable market; while for the home garden and market a longer season is desirable.

FIG. 33 – ATLANTIC PRIZE, AND ITS NORMAL FOLIAGE

Foliage.—Abundant, broad and close, or scanty cut and open foliage is sometimes of importance, according to whether the location, season and other conditions make it desirable that the foliage protect the fruit from the sun or admit the sunlight, with as little obstruction as possible, to the center of the plant. [Pg 102] In different sorts, we have gradations from those in which the leaves are so deeply cut as to have a fern-like appearance, to those like the Magnus, or potato-leaved, in which the margin of each leaflet is entire, and from those in which the leaflets are so few and small as to scarcely shut out the light at all to those in which they are so numerous that the light can hardly penetrate to the center of the plant. The Atlantic Prize is an illustration of the scanty foliaged sorts, and the Royal Red or Buckeye State of those in which it is more abundant. As to color, the foliage varies from the dark blue-green of the Buckeye State to the light, distinctly yellowish-green of the Honor Bright.

Varietal differences as to fruit.—These are often more important than those of vine. For canning, for forcing, and some other uses and for certain markets, a medium and uniform size is a very important quality, while in other cases uniformity is not important and the larger the individual fruits, provided they be well formed, the better. We have different sorts in which the size of the fruit varies from that of the Currant, which is scarcely 1 inch in circumference, to that of Ponderosa, of which well-formed specimens over 20 inches in circumference have been grown.

[Pg 103]

FIG. 34—DWARF CHAMPION. NOTE CHARACTER OF FOLIAGE

Shape.—It is always desirable that the outline of the vertical section shall be a flowing line with a broad and shallow, or no depression at the stem end and as little as possible at the opposite point; but the relative importance of this, or whether the general outline shall be round or oval, either vertically or horizontally, forming a round, long or flat fruit, is largely determined by how the fruit is to be used, and by in [Pg 104] dividual taste. A round fruit is best for canning; a long one is the most economical for slicing, though some prefer a flat one for this purpose. It is always desirable that the outline of the horizontal section shall be smooth, flowing and symmetrical, and if there be any distinct sutures that they shall be shallow and broad; but the relative importance of this, and whether the outline be round or oval, is wholly a matter of individual taste. Some people and markets prefer one shape and others a very different one. Size and smoothness of fruit are the factors which control price in some markets, while in others these points are quite secondary to color and character of flesh.

We have sorts which vary from the perfectly spherical ones of the grape and cherry, to those in which the vertical diameter is less than a third of that of the horizontal section; and the pear-shaped in which the vertical diameter is twice or thrice that of the longest horizontal section, and from those in which the outline of both the vertical and horizontal sections is smooth and flowing to those in which the vertical section has a deep indentation at both the stem and opposite ends, and those in which the horizontal section is broken by deep indentures and sutures often disposed with great irregularity.

For shipping long distances, for the rough handling, and for the easy preparation for the fruit for canning, a thick, tough skin is desirable, while for home use it is objectionable. Freedom from blemish or skin crack is also often an important quality, and we have sorts which vary greatly in these respects. The color of the skin, whether purple, red, yellow or white, is [Pg 105] a matter of taste. In some markets the choice is given to purple fruit, like the Beauty, while in others it can only be sold at a reduced price. There are few who would care to use either yellow or white fruit for canning or cooking in any way, but many prefer them for slicing, or like to use them with the red for this purpose; we have sorts showing every gradation from white or light yellow in color through shades of red to dark purple-red, and still others which show distinct colors in splashings and shadings.

Character of flesh.—Many consider that the greater the number of cells and the larger the proportion of flesh to that of pulp and seed the better. This may be true of itself, but the fruit-like acid tomato flavor which most people value is found chiefly in the pulp, and the fruit which has not a due proportion of pulp and flesh seems to be insipid and tasteless. Again, the division into many small cells is often connected with a large and pithy placenta and unevenness in maturity and coloring, which faults often more than overbalance any advantage from small cells and thick flesh. The size and character of the placenta are important qualities.

In some sorts it is large, dry, pithy and hard, extending far into the fruit even to below the center; and sometimes seems to divide into secondary or branch placentas or masses of hard cellular matter, while in other varieties it is small and so soft and juicy as scarcely to be distinguished from the flesh. Usually, but not invariably, the large and pithy placenta is correlated with large-sized fruit having many cells; where this is the case it practically necessitates the [Pg 106] cutting away and wasting of a large proportion of the fruit in preparing it for canning, so that the canners usually prefer round, medium-sized fruits.

The character of the interior of the fruit varies greatly in different varieties. Both the exterior and divisional walls vary in thickness and in consistency. In some varieties they are comparatively thin, hard and dry; in others, thicker, softer and more juicy. In some cases there is but little interior wall, the fruit being divided into but few—even but two—cells of even size and shape, while in others there are many cells of varying size and shape. Varieties also differ greatly as to the amount, consistency and flavor of the pulp and the number of seeds. It requires from 300 to 500 pounds of ripe fruit to furnish a pound of seed of Ponderosa, while with some of the smaller, earlier sorts one can get a pound of seed from 100 to 200 pounds of fruit.

Coloring and ripening.—Uniformity and evenness in coloring and ripening are an important quality. Tomatoes generally color and ripen from within outward, and from the point opposite the stem upward, but varieties differ in the evenness and rapidity with which this takes place. It is always desirable that the ripening be as even as possible and that there be no green and hard spots either at the surface or in the flesh, but often perfection in this respect is correlated with such lack of size and solidity as to counterbalance it. Rapidity in ripening, in a general way, is desirable for fruit to be used at home, and undesirable in that which is to be shipped.

The time a tomato fruit will remain in usable con [Pg 107] dition and the amount of rough handling it will endure without becoming unsalable are most important commercial qualities depending largely upon the combined effects of the form and structure of the fruit, solidity and firmness of the flesh and ripening habit. In all these respects we have varieties which differ greatly, from the Honor Bright, which requires as much time to ripen, and when ripe is firm-fleshed and will remain usable as long as a peach, to those which 24 hours after reaching their full size are fully colored and ripe, and in 24 hours more are so over-ripe and soft that they will break open of their own weight.

These are only some of the varietal differences of the tomato. Are such differences of practical importance? I think they are, and that a wise selection of the type best suited to one's own particular conditions and requirements is one of the most essential requisites of satisfactory tomato culture. How important it seems to practical tomato growers may be illustrated by an actual case.

In a certain section of New Jersey the money-making crop is early tomatoes, and they are grown to such an extent that from an area with a radius of not exceeding 5 miles they have shipped as much as 15,000 bushels in one day, and the shipments will often average 8,000 bushels for days together. They have tried a great number of sorts, but have settled upon a certain type of a well-known variety as that best suited to their conditions and needs. Seeds of this variety which are supposed to produce plants of the exact type wanted can be bought from seedsmen for 10 cents an ounce and at much lower rates for [Pg 108] larger quantities, but when one of the most successful growers of that locality, because of change of occupation, offered seed selected by him for his own use for sale at auction, it brought $3 an ounce. This price was paid because of the confidence of the bidders that the seed could be depended upon to produce plants of the exact type wanted for their conditions; and I was assured that the use of this high-priced seed actually added very largely to the profits from every field in that vicinity in which it was used, but the use of some of the same lot of seed by planters in Florida resulted in financial loss because the type of plant produced was not suited to their conditions and requirements.

A wise answer can only be given after a study of each case, and no one can do this so well as the planter himself. He should know, as no one else can know, his own conditions and requirements, and should be able to form very exact ideas of just what he wants, and the doing so is, in my opinion, one of the most important requisites for satisfactory tomato growing. I also believe that it is as impossible for a man to answer offhand the question, "What is the best variety of tomato?" as for a wise physician to answer the question, "What is the best medicine?"

Varietal names and descriptions mean something quite different in the case of plants like the tomato, which are propagated by seed, from what they do with plants like the apple and strawberry, which are propagated by division. In the latter case all the plants of the variety are but parts of the primal origination, and so are alike. A description is simply a more or [Pg 109] less complete and accurate definition of what a certain immutable thing really is, but in the case of plants propagated by seed the variety is made up of all the plants which accord with a certain ideal. Bailey says, "Of all those which have more points of resemblance than of difference," and a description of the variety is of that ideal which in common practice is not fixed, but may and generally does vary not only with different people but from time to time. The only foundation for varietal names in plants of this class is an agreement as to the ideal the name shall stand for. Under modern horticultural practice when anyone has been able to secure seed most of which he is reasonably sure will develop into plants of a distinct type different from that of any sort known to him, he has a distinct variety, so that it is not surprising that we should find that American seedsmen offer tomato seed under more than 300 different names, and those of Europe under more than 200 additional, so that we have more than 500 varietal names, each claiming to stand for a distinct sort. Now it is quite possible—indeed, it is certain—that we might have 500 tomato plants each different in some respect, either of vine, leaf, habit of growth, or character of fruit, from any of the others and that these differences might make plants of one type better suited to certain conditions and uses than any other; but it is very certain that these 500 names do not stand for such differences. It is doubtless true that a portion—though I think but a small portion—of these different sorts exist simply as a matter of commercial expediency; but by far a greater part of them exist because one has found that [Pg 110] plants of a certain character were better suited to some set of conditions and requirements than any sort with which he was acquainted, and having secured seed which he thought would produce plants of that character, has offered it as of a distinct sort.

It is probable that a better acquaintance with sorts already in cultivation would have prevented the naming of many of these stocks as distinct varieties. What is of far more practical importance, the same name does not always stand for precisely the same type with different seedsmen, or even with the same seedsmen in different years; nor are the seedsmen's published descriptions such as would enable any one to learn from them just what type he will receive under any particular name, or which sort he should buy in order to get plants of any desired type. Seedsmen's catalogs are published and distributed gratuitously at great expense, and are issued, primarily, for the sake of selling the seeds they offer. They answer the purpose for which they are designed, in proportion as they secure orders for seeds. Will this be measured by the accuracy and completeness of their descriptions? I think that it needs but slight acquaintance with the actual results of advertising to answer in the negative, and whatever your answer may be, the answer given by the catalogs themselves is an emphatic no.

In a recent case I looked very carefully through the catalogs of 125 American seedsmen who listed a certain variety which is very markedly deficient in a certain desirable quality, and found that but 37 of the 125 mentioned the quality in connection with the variety at all and of these but 7 admitted the defi [Pg 111] ciency, while 30 told the opposite of the truth. Even if a complete, exact and reliable description of a variety was published by disinterested persons, one could not be sure of getting seed from seedsmen which would produce plants of that exact type, since there is no agreement or uniformity among them as to the exact type any varietal name shall stand for.

One way of getting seed of the exact type wanted is to do as the South Jersey growers did: go to work and breed up a stock which is uniformly of the type wanted; but this involves more painstaking care than many are willing to give, though I think not more than it would be most profitable for them to expend for the sake of getting seed just suited to their needs.

A second and easier way is to secure samples of the most promising sorts and from the most reliable sources and grow them on one's own farm; select the stock which seems best for him and buy enough of that exact stock for several years' planting, and in the meantime be looking for a still better one. Tomato seed stored in a cool, dry place will retain its vitality for from three to seven years.

[Pg 112]

CHAPTER XVI

Seed Breeding and Growing

The potentialities of every plant and its limitations are inherent, fixed and immutable in the seed from which it is developed and are made up of the balanced sum of the different tendencies it receives in varying degree from each of its ancestors back for an indefinite number of generations. A very slight difference in the character or the degree of any one of the tendencies which go to make up this sum may make a most material difference in the balance and so in the resulting character of the plant produced. Different plants, even of the same ancestry, vary greatly in prepotency or in the relative dominance of the influence they have over descendants raised from seed produced by them.

In some cases all the plants raised from seed produced by a certain plant will be essentially alike and closely resemble the seed-bearing plant, while seed from another plant of the same parentage will develop into plants differing from each other and seemingly more influenced by some distant ancestor or by varying combinations of such influences than of those of the plant which actually produced the seed from which they were developed. Successful seed breeding can only be accomplished by taking advantage of these principles of heredity and variation, and by a wise use of them it is possible to produce seed which can [Pg 113] be depended upon to produce plants of any type possible to the species.

The first essential for breeding is to have a clear and exact conception of precisely what, in all respects, the type shall be and then the securing of seed which has come from plants of that exact character for the greatest possible number of generations, carefully avoiding the introduction by cross-pollination of tendencies from plants differing in any degree from the desired type. Secondly, seed should be used from plants which have been proven to produce seed, which will develop into plants like themselves or are strongly prepotent. A practical way to accomplish this in the tomato is as follows:

By experiment and observation form a very clear conception of precisely the type of plant and fruits which is best suited to your needs. This may be done by the study of available descriptions of sorts, by conference with those who have had experience in your own or similar climatic and soil conditions and in raising fruit for the same purposes and, best of all, by trials of samples of different sorts and stocks on your own grounds. Having formed such a conception, write out the clearest possible description of exactly what you want and the ideal plant you are aiming at, stating as fully and minutely as possible every desirable quality and also those to be avoided. I consider not only the formation of an exact ideal, but the writing out of a most minute and exact description of precisely what in every particular the ideal plant should be and the rigid adherence to that exact ideal in selection, as the most important ele [Pg 114] ments of successful seed breeding. Without it one is certain to vary from year to year in the type selected and in just so far as he does this, even if it be toward what might be called improvements or in regard to an unimportant quality, he undermines all his work and makes it impossible to establish a strain which can be relied upon to produce an exact type.

With this description in hand, search out one or more plants which seem the nearest to the ideal. In doing this it should be kept in mind that the character of the seed is determined by the plant rather than by the individual fruit. Therefore, a plant whose fruit is most uniformly of the desired type should be chosen over one having a small proportion of its fruits of very perfect type, the others being different and variable. Save seed from one or more fruits from each of the selected plants, keeping that from each fruit, or at least each plant, separate. Give it a number and make a record of how nearly, in each particular, the plant and fruit of each number come to the desired ideal. I regard the saving of each lot separately and recording its characters as very important, even when all have been selected to and come equally close to precisely the same ideal. Quite often the seed of one plant will produce plants precisely like it, while that of another, equal or superior, will produce plants of which no two are alike and none like that which produced the seed, so that often the mixing of seed from different plants of the same general type, and seemingly of equal quality, prevents the establishment of a uniform type.

The next year from 10 to 100 plants raised from [Pg 115] each lot are set in blocks and labeled. As they develop the blocks are studied and compared with the original description of the desired type and that of each plant from which seed was saved, and the block selected in which all the plants come the nearest to the desired type, and which show the least variation. From it plants are selected in the same way and to the same type as the previous year. It is better to make selections from such a block than to take the most superior plants from all of the blocks, or from one which produced but one or but a few superlative ones, the rest being variable.

It is also well to consider the relative importance of different qualities in connection with the degree to which the different lots approach the ideal in these respects. Such a course of selection intelligently and carefully carried out will give, in from three to five years, strains of seed greatly superior and better adapted to one's own conditions than any which it is possible to purchase. A single or but a very few selections may be made each year, and the superior value of the seed of the remainder of the seed blocks for use in the field will be far more than the cost of the whole work.

Growing and saving commercial seed.—The ideal way is for the seedsman to grow and select seed as described above and give this stock seed to farmers who plant in fields and cultivate it, much as is recommended for canning, and save seed from the entire crop, the pulp being thrown away. Only a few pickings are necessary and the seed is separated by machines worked by horse power at small cost, often not [Pg 116] exceeding 10 cents a pound. They secure from 75 to 250 pounds per acre, according to the variety and crop, and the seedsmen pay them 40 cents to $1 a pound for it. Some of our more careful seedsmen produce all the seed they use in this way; others buy of professional seed growers, who use more or less carefully grown stock seed. In other cases when the fruit is fully ripe it is gathered, and the seeds, pulp and skins, are separated by machinery; the seed is sold to seedsmen, the pulp made into catsup, and only the skins are thrown away. Still others get their supply by washing out and saving the seed from the waste of canneries. Such seed is just as good as seed saved *from the same grade of tomatoes* in any other way, but the fruit used by the canneries is, usually, a mixture of different crops and grades, and even of different varieties, and consequently the seed is mixed and entirely lacking in uniformity and distinctness of type.

Generally from 5 to 20 per cent. of the plants produced by seed as commonly grown either by the farmer himself or the seedsmen, though they may be alike in more conspicuous characteristics, will show varietal differences of such importance as to affect more or less materially the value of the plant for the conditions and the purposes for which it is grown. In a book like this it is useless to attempt to give long varietal descriptions even of the sorts commonly listed by seedsmen, since such descriptions would be more a statement of what the writer thought seed of that variety should be rather than of what one would be likely to receive under that name.

[Pg 117]

CHAPTER XVII

Production for Canning

Growing for canning has many advantages over growing for market. Some of these are that it is not necessary to start the plants so early, that they can be grown at less cost, and set in the field when smaller and with less check, and on this latter account are apt to give a large yield. It is not necessary to gather the fruit so often, nor to handle it so carefully, while practically all of it is saleable. For these reasons the cost of production is lower and it is less variable than with crops grown for market. Still farmers and writers do not agree at all as to the actual cost. It is claimed by some that where the factory is within easy reach of the field the cost of growing, gathering and delivering a full yield of tomatoes need not exceed $12 to $18 an acre, while others declare they cannot be grown for less than $40. Nearly one-third of this cost is for picking and delivering, and varies more with the facilities for doing this easily and promptly and with the yield than with crops grown for market. A large proportion of the crops grown for canning are poorly cultivated and unwisely handled, so that the average yield throughout the entire country is very low, hardly exceeding 100 bushels an acre. But where weather and other conditions are favorable, and with judicious cultivation, a yield of [Pg 118] 300 to 800 bushels an acre can be expected. I have known of many larger ones.

A large proportion of the tomatoes grown for canning are planted under contract, by which the farmer agrees to deliver the entire yield of fruit fit for canning, which may be produced on a given area, at the contract price per bushel or ton. The canner is to judge what fruit is fit for canning and this often results in great dissatisfaction. To the grower it seems in many cases as though the quantity of acceptable fruit paid for was determined quite as much by the abundance or scarcity of the general crop as by the weight hauled to the factory. The prices paid by the factories for the past 10 years run from 10 to 25 cents a bushel, while canning tomatoes in the open market for the same period have brought from 8 to 50 cents a bushel, which, however, are exceptional prices. In all but two of the past 10 years uncontracted tomatoes could generally be sold, in most sections, for more than was paid on contract. I have given the price a bushel, though canning tomatoes are usually sold by the ton. The cost of the product of a well-equipped cannery is divided about as follows: fruit, 30 per cent.; handling, preparing and processing the fruit, 18 per cent.; cost of cans, labels, cases, etc., 43 per cent.; labeling, packing and selling, 0.035 per cent.; incidentals, 0.055 per cent.

Canning on the farm.—While as a general proposition such work as canning tomatoes can usually be done at less cost in a central plant, yet in many cases the grower can profitably do this on the farm, thus saving not only the expense of delivery at the factory, [Pg 119] but the dissatisfaction with weights credited and delays in receiving the fruit. But very little special apparatus or machinery (more than some form of boiler for supplying steam) is needed, and this and the cans can be readily obtained of dealers in canners' supplies. In Maryland and neighboring states many dealers furnish all necessary machinery, cans and other requisites and contract for the crop delivered in cans.

Canning on the farm where the fruit is grown would be more generally practiced except for the popular demand that the canned product shall be brighter colored than it is possible to produce from fruit alone, and the necessary dyeing and other doctoring can be more easily and skilfully done at a central factory, though it is always at the sacrifice of flavor and healthfulness for the sake of appearance. Another advantage of canning on the farm is that it can be done with less waste of fruit. The hauling to the factory and delay in working the fruit result in a great deal of waste. The average cannery does not obtain more than 1,200 pounds of product from a ton of fruit, there being 800 pounds of waste, while with sound, ripe, perfectly fresh fruit, it is entirely practical to secure from 1,600 to 1,800 pounds of canned goods from a ton, and this saving in waste would more than counterbalance the gain from the use of the better machinery possible in the factory.

The process of canning is simple and consists first of rinsing off the fruit, then in wire baskets or pails dipping it into boiling hot water to start the skins, which will require but two to four minutes. While they are still hot they should be peeled and imperfec [Pg 120] tions cut out, then promptly placed in the cans, which should be fully filled; it is well to do this by adding the juice which has escaped while peeling, instead of water, as is done in the larger factories. This will give the canned fruit better color and lessen the need of dye. Place in a hot box for three to five minutes until heated through, wipe top of can clean and drop perforated cap in place, add flux and solder, seal cap in place with round capper, close perforation in cap with drop of solder. Place in box or kettle and steam or boil for 20 to 40 minutes. If the tomatoes were all ripe and none over-ripe, and have been kept hot from the time they went into the scalding kettle until the sealed cans are in the kettle, 20 minutes' cooking will make them surer to keep than 40 minutes would with fruit such as is commonly received at factories, or that which has been allowed to cool once or twice while in process.

[Pg 121]

CHAPTER XVIII

Cost of Production

There are a few vegetables or fruits where the cost of production and the price received are more variable than with the tomato. The cost per acre for raising the fruit varies with the conditions of soil, facilities for doing the work economically and with the season, while that of marketing the product varies still more. Under usual conditions, the growing of an acre of tomatoes and the gathering and marketing of the fruit will cost from $18 to $90, of which from 15 to 40 per cent. is spent in fertilizing and preparing the ground, 5 to 10 per cent. for plants, 20 to 30 per cent. for cultivation, and 25 to 40 per cent. for gathering and handling the fruit. The last item, of course, varies somewhat with, but not in proportion to, the amount of the crop, as it costs proportionately less to gather a large than a small crop, and for canners' use than for market.

The expense of shipping and marketing the crop varies so greatly according to the conditions and methods that I do not attempt to state the amount. The total yield of fruit runs from 200 to 600 or 700 bushels to the acre, a 200-bushel crop of tomatoes comparing as to amount with one of 25 bushels of wheat and a 700-bushel crop of tomatoes with one of 60 bushels of wheat; with the best and wisest cultivation and under the most favorable conditions one can as [Pg 122] reasonably hope for one as for the other. Of this total yield, from 10 to 25 per cent. of the fruit should be such as, because of earliness and quality, can be sold as extras, and there is usually from 5 to 10 per cent., and sometimes a much larger per cent., which should be rejected as unsalable. The selected fruit should net from $1 to $5 a bushel, the common from 30 to 75 cents—making the returns for a 200-bushel yield well sold in a nearby market $70 to $350, and proportionately larger, for a better yield. In practice I have known of crops which gave a profit above expenses of over $1,000 an acre. This came, however, from exceptionally favorable conditions and skilled marketing, and I have known of many more crops where, though the fruit was equally large and well grown, the profit was less than $100.

In this country a greenhouse is seldom used solely for the growing of tomatoes, but other crops—such as lettuce—are grown in connection with the tomatoes, so that it is impracticable to give the cost of production. As grown at the Ohio state experiment station—and the crop ripened in late spring or early summer and sold on the market of smaller cities—greenhouse tomatoes have yielded about two pounds a square foot of glass and brought an average price of 12 cents per pound. In other cases yields as high as 10 pounds a foot of glass and an average price of 40 cents a pound have been reported.

[Pg 123]

CHAPTER XIX

Insects Injurious to the Tomato
By Dr. F. H. Chittenden
Bureau of Entomology, U. S. Department of Agriculture

From the time tomato plants are set in the field until the fruit has ripened they are subject to the attacks of insects which frequently cause serious injury. On the whole, however, the tomato is not so susceptible to damage as are some related crops—such as the potato.

Cutworms.—Of insects most to be feared and of those which attack the plants when they are first set out are cutworms of various species. The grower is as a rule quite too familiar with these insects, and no description of their methods is necessary, beyond the statement that they cut off and destroy more than they eat and re-setting is frequently necessary. The best remedy is a poisoned bait, prepared by dipping bunches of clover, weeds, or other vegetation in a solution of Paris green or other arsenical, 1 pound to 100 gallons of water. These baits are distributed in small lots over the ground *before* the plants are set, the precaution being observed that the land is free for two or three weeks from any form of vegetation. This will force the hungry "worms" to feed on the baits, to their prompt destruction. A bran-mash is also used instead of weeds or clover, and is prepared [Pg 124] by combining one part by weight of arsenic, one of sugar, and six of sweetened bran, with enough water added to make a mash. The baits are renewed if they become too dry, or they can be kept moist by placing them under shingles or pieces of board.

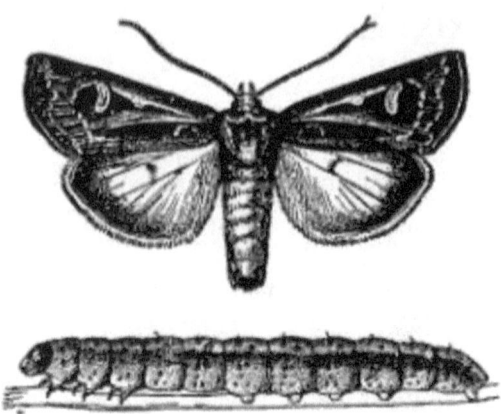

FIG. 35 — CUTWORM AND PARENT MOTH (*Feltia subgothica*) (From Chittenden, U. S. Department of Agriculture)

Flea-beetles attack the plants soon after they are set, and their injuries can be prevented by dipping the young plants before setting in a solution of arsenate of lead, about 1 pound to 50 gallons of water, or Paris green, 1 pound to 100 gallons. If this precaution has not been observed a spray of either of these arsenicals used in the proportion specified will suffice, repeating if the insects continue on the plants. In the preparation of the spray a pound of fresh lime to each pound of the arsenical should be added; or, better yet, Bordeaux mixture should be employed as a diluent instead of water. This mixture has some insecticidal value, is a most valuable fungicide, and [Pg 125] is also a powerful deterrent of flea-beetle attack, acting to a less degree against other insects which are apt to be found on the tomato. In applying any spray a sprayer costing not less than $7 is a positive necessity.

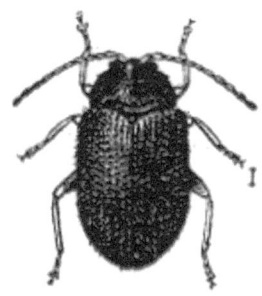

FIG. 36 — FLEA-BEETLE
Does great injury to young plants. Much enlarged. Actual size shown by line at right. (From Chittenden)

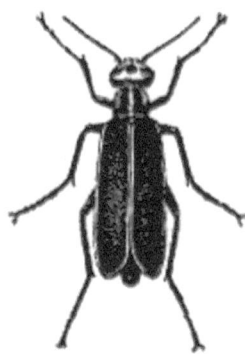

FIG. 37 — MARGINED BLISTER BEETLE

The Colorado potato beetle, or "potato bug," sometimes injures tomatoes, but not as a rule when potatoes are available. This suggests the use of potatoes as a trap crop, planted in about three rows completely around the field of tomatoes. The arsenicals used in the same proportion as for flea-beetles will destroy the potato beetle. It is necessary to keep the trap potatoes well sprayed to prevent them from breeding on these plants and migrating to the tomatoes. Potato beetles can also be controlled by jarring them from the affected plants into large pans containing a little water on which a thin scum of kerosene is floating.

Blister beetles may be controlled, under ordinary [Pg 126] circumstances, by the same method employed against the Colorado beetle. When they are present in great numbers a good remedy consists in driving them with the wind from the cultivated fields into windrows of straw or similar dry material previously prepared along the leeward side of the field, where they will congregate and can be burned.

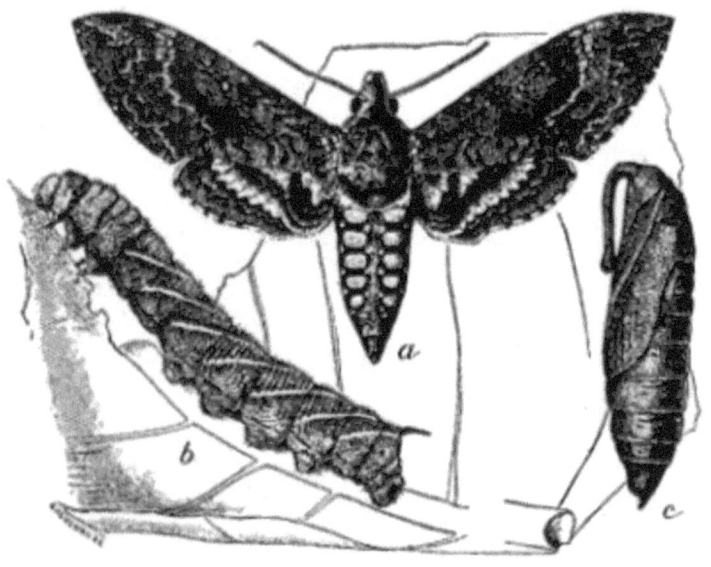

FIG. 38 — TOMATO WORM (*Protoparce sexta*)
(*a*) Adult moth; (*b*) full-grown larva; (*c*) pupa — all reduced. (After Howard, U. S. Dept. Agr.)

The tomato worms, of which there are two common species closely resembling each other, are often abundant and destructive on tomato foliage, particularly southward. The arsenicals will kill them, or they can be held in check by hand-picking, a little [Pg 127] experience enabling one to detect their presence readily. Turkeys are utilized in destroying these worms in the South.

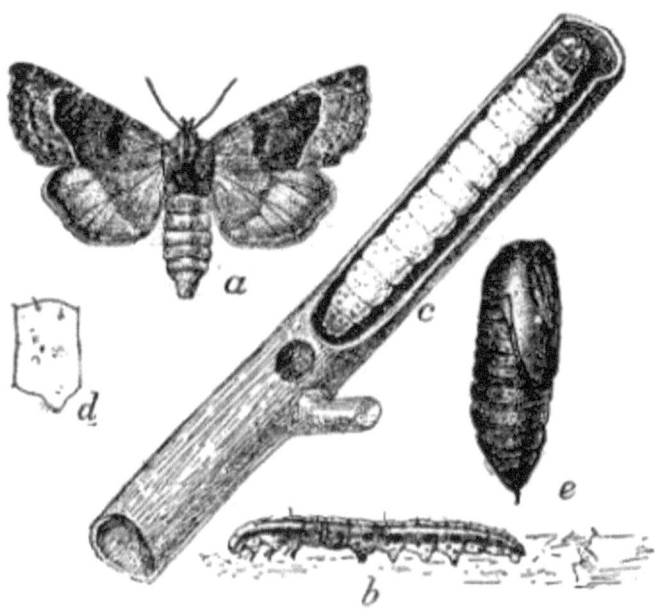

FIG. 39 — TOMATO STALK-BORER (*Papaipema nitela*)
(*a*) Female moth; (*b*) half-grown larva; (*c*) mature larva in injured stalk; (*d*) lateral view of abdominal segment; (*e*) pupa — all somewhat enlarged. (From Chittenden, U. S. Dept. Agr.)

The stalk-borer, as its name implies, attacks the stalk, and is an intermittent pest, though quite annoying at times. It is difficult to combat, but its injuries may be prevented by care in keeping down, and by *promptly* destroying, the weeds after they are pulled or hoed out during the growing season. If weeds are left to dry the striped caterpillar of this species [Pg 128] will desert them and enter cultivated plants. Ragweed and burdock are the principal foods of this insect, and special attention should be given to eradicate them where tomatoes are planted. Crop rotation is advisable where this can be conveniently practiced, and such plants as cabbage, radish and the like, onions, beets, asparagus and celery are suggested as alternates. When the plants are sprayed with arsenicals for other insects this will operate to a certain extent against the stalk-borer.

FIG. 40 — CHARACTERISTIC WORK OF THE TOMATO FRUIT WORM (*Heliothis obsoleta*)
(Redrawn by Johnson from C. V. Riley)

FIG. 41 — ADULT MOTH, OR PARENT OF TOMATO FRUIT WORM
(From Chittenden, U. S. Department of Agriculture)

The tomato fruit worm (Fig. 40) known as the bollworm of cotton and the ear worm of corn, is frequently the cause of serious trouble to tomato growers, especially in the southern states, due to its pernicious habit of eating into and destroying the green and ripening fruit. For its control it is advisable not to [Pg 129] plant tomatoes in proximity to old corn or cotton fields, nor should land be used in regions where this species is abundant until it has been fall or winter plowed. Sweet corn planted about the field before the tomatoes are set will serve as a lure for the parent moths to deposit their eggs, corn and cotton being favorite foods of this species and preferred to tomatoes. The fruit worm feeds to a certain extent on the foliage before penetrating the fruit, and it is possible to keep it in subjection by spraying with arsenicals as advised for the flea-beetles. It is suggested that arsenate of lead, being more adhesive than other arsenicals, should be used for the first sprayings, beginning when the fruit commences to form, repeating once or twice as found necessary, and making a last spraying with Paris green within a few days of ripening. This last poison will readily wash off and there is no danger whatever of poisoning to human beings, as has been conclusively proved in numerous similar cases. For the perfect success of this remedy the [Pg 130] last spraying is essential, as those who have sprayed with an arsenical and have reported only partial good results have discontinued within about two weeks of the time of the ripening of the first fruit.

White fly or aleyrodes.—These minute insects are familiar to most growers who raise tomatoes under glass. They can be held in control by vaporization or fumigation with tobacco or nicotine extracts, or by spraying with kerosene emulsion or the so-called whale-oil (fish-oil) soap. Care is necessary in using the extracts that the smudge does not become too dense and injure the plants. Before applying this remedy on a large scale a preliminary trial should be made following the directions on the packages, and reducing the amount if any ill results follow. Hydrocyanic acid gas, properly used, is also an excellent remedy for aleyrodes, aphides, "mealybug," and other soft-bodied insects which are sometimes troublesome on greenhouse tomatoes.

For a complete account of the methods of making and handling hydrocyanic acid gas, see Professor Johnson's book entitled "Fumigation Methods," published by Orange Judd Company, of New York. Sent postpaid for $1.—[Author.

[Pg 131]

CHAPTER XX

Tomato Diseases

By W. A. Orton

U. S. Department of Agriculture

DISEASES NOT CAUSED BY FUNGI OR INSECTS

The health of tomato plants is to a large extent dependent on the conditions under which they are being grown. The character and physical condition of the soil, the supply of water and plant food, the temperature and amount of sunlight, are all factors of the greatest importance in the growth and development of the crop. When there are variations from the normal in the case of any of these the plant adapts itself to the change as far as possible, but its functions may be so disturbed as to result in ill health or disease. It is in many cases difficult to draw a line between a natural re-action of the plant to its environment and a state of disease. For example, the trouble described in the next paragraph seems to fall into the first class.

Shedding of blossoms.—The tomato is very liable to drop its buds and blossoms and in some instances partial or total crop failures have resulted. The principal causes are an over-rapid growth, due in many cases to an excess of nitrogenous fertilizers, unfavorable weather conditions, especially cold winds, continued rainy or moist weather, which hinders pollina [Pg 132] tion, lack of sunlight, or extremely hot weather. Such shedding can be partially controlled by pruning away the lateral branches as soon as formed and topping the plants after the third cluster of fruit has set, and by a reduction in the use of nitrogenous fertilizers. A failure to set fruit in the greenhouse is often due to lack of pollination, which must be remedied by hand pollination.

Cracking of the fruit.—The formation of cracks or fissures in the nearly mature fruit is due to variations in the water supply and other conditions affecting growth at this stage. If after the development of the outer portion of the fruit has been checked by drought there follows a period of abundant water supply and rapid growth, the fruit expands more rapidly than its epidermis and the latter is ruptured. Some varieties of tomatoes are much less subject to this trouble than others and should be given preference on this account. The grower, so far as lies in his power, should seek to maintain an uninterrupted growth by thorough preparation of the land, by cultivation or by mulching. If the half-grown fruits are enclosed in paper bags, cracking may be prevented, but at the risk of loss of flavor in the ripened fruit.

Leaf curl.—The effect of pruning is to stimulate growth and to increase the size of the leaves, the effort of the plant being to maintain a balance between roots and foliage. With rapidly growing plants, especially in the greenhouse and garden where both high manuring and pruning have been practiced, more or less curling and distortion of the leaves may result without developing into serious trouble if the grower [Pg 133] takes the hint and modifies his methods so as to permit a more balanced growth. On the other hand, the ill effects of over-feeding and pruning may reach a point where the plant is seriously crippled.

Edema.—Under certain conditions plants in greenhouses or even in the open field, may absorb water through the roots faster than it can be transpired through the leaves, with the result that dropsical swellings or blisters occur on the leaves and more succulent stems. There is also a deformation of the foliage, much like the leaf-curl produced by over-feeding. This trouble, known as edema, occurs when the soil is warmer than the air, or during periods of moist, warm, cloudy weather, which checks transpiration. The grower should cease pruning, and withhold water, and in the field cultivate deeply. In the greenhouse, adequate ventilation should be given, keeping the house dry rather than moist.

Mosaic disease.—The tomato is occasionally subject to a trouble allied to the mosaic disease of tobacco. It is characterized by a variegation of the leaves into light and dark green areas, usually accompanied by distortion and reduction in size. In severe cases a whole field may become worthless. While the nature of this malady is not fully understood, it is known to be due to a disordered nutrition of the young leaf-cells. It can be produced by severe pruning or by mutilation of the roots in transplanting, both of which should be carefully avoided. It is more likely to occur in seedlings that have made a soft, rapid growth on account of an excess of nitrogenous fertilizer or too high temperature. Close, clayey soils, on account of their [Pg 134] poor physical condition, also favor the development of the disease after transplanting.

Western blight (Yellows).—In the North Pacific and Rocky Mountain states, serious losses are annually caused by a disease apparently unlike any eastern trouble. It is marked by a gradual yellowing of the foliage and fruit. Development is checked, the leaves curl upward and the plant dies without wilting. The nature and cause of this disease is as yet unknown. It appears to be worst on new land. Experiments that have been made indicate that in older cultivated fields thorough preparation of the soil, manuring and cultivation, combined with care in transplanting to avoid injuring the roots and checking growth, will greatly restrict the spread of this blight.

DISEASES CAUSED BY PARASITES

There are several fungous parasites of tomatoes, which, for the readers convenience, may be briefly mentioned and the treatment of all discussed together. The first three are indeed somewhat difficult to tell apart without a microscope, as they produce a similar effect on the leaves and all yield to the same treatment—thorough spraying with Bordeaux mixture.

Leaf spot (*Septoria lycopersici* Speg.) has been widely prevalent and injurious during recent years. It produces small, roundish dark-brown spots on leaves and stems. The lower leaves are attacked first and gradually curl up, die and fall off. The vitality of the plant is reduced and it is only kept alive by the young leaves formed at the top. [Pg 135]

The fungus that causes early blight of potatoes (*Alternaria solani* (E. & M.) J. & G.) occurs on tomatoes also, sometimes doing much injury. The spots formed are at first small and black, later enlarging and exhibiting fine concentric rings.

A somewhat similar leaf-blight results from a species of *Cylindrosporium*, and other fungi may occur on diseased leaves.

Leaf mold (*Cladosporium fulvum* Cke.) is quite distinct from the foregoing in appearance. It does not cause such distinct spots but occurs in greenish brown, velvety patches of irregular outline on the under side of the leaves. The lower leaves are first attacked, and as the disease progresses they turn yellow and drop off. This is the principal fungous enemy of greenhouse tomatoes, but also does injury in gardens, particularly in Florida and the Gulf region. It is readily controlled by spraying. In the greenhouse care should be taken to ventilate well, without, however, allowing cold drafts to strike the plants.

Downy mildew (*Phytopthora infestans* DeBy.), the cause of the late blight of potatoes, will attack tomatoes during cool and very moist weather, which greatly favors its development. Such outbreaks sometimes occur to a limited extent in New England and serious losses are reported on the winter crop in southern California, but the disease has never been troublesome in other sections of the country, as it cannot develop in dry or hot weather. It affects the tomato as it does the potato, forming on the leaves dark, discolored spots, which spread rapidly under favorable conditions, killing the foliage in a few days. The fruit is [Pg 136] also attacked and becomes covered with the mildew-like spore-bearing threads of the fungus. Bordeaux mixture properly applied is an efficient preventive.

Spraying tomatoes.—It should be the invariable practice of the tomato grower to spray with Bordeaux mixture to prevent injury from any of these leaf-blights. This should be done while the plants are still healthy, as if put off until the disease appears the battle is half lost. Make the first application to the young plants in the seed-bed a few days before transplanting. Spray again within a week after the plants are set in the field, and repeat at intervals of ten days or two weeks until the fruit is full grown. Success in spraying depends mainly on the thoroughness of the work. The aim should be to cover every leaf with a fine mist. Do not drench the foliage but pass to the next plant before the drops run together and off the leaf. Use a nozzle that gives a fine spray and maintain a high pressure at the pump.

Preparation of Bordeaux mixture.—Formula: Copper sulphate (bluestone), 5 pounds; lime, 5 pounds; and water, 50 gallons. The copper sulphate may be either in crystals or pulverized. Dissolve by suspending the required amount in a coarse sack near the top of the water a few hours before it will be needed. The lime must be fresh stone lime of good quality. Slake thoroughly by the addition of small quantities of water at a time as needed, stirring until all small lumps are slaked. Strain both the lime milk and the copper sulphate or bluestone solution through a brass strainer of 18 meshes per inch and dilute each with half the water before mixing together. Do not use [Pg 137] Bordeaux left over from the previous day. An old mixture or one made from the concentrated solutions has a poor physical condition. It settles more quickly, tends to clog the nozzle and does not adhere so well to the foliage. Failure to use the strainer results in endless trouble in the field from clogged nozzles.

FIG. 42—PROPER WAY TO MAKE BORDEAUX
(From W. G. Johnson)

When much spraying is to be done it is more convenient to keep the bluestone and lime in separate permanent stock solutions, as shown in Fig. 42, containing 2 pounds to the gallon of their respective ingredients. These will keep indefinitely, if the water evaporated is replaced, and may be used from as needed.

Spraying apparatus.—Tomato growers having only a small area to spray may use one of the numerous forms of hand-pumps or bucket sprayers now on the [Pg 138] market. For larger fields it will be necessary to employ a barrel sprayer. This consists of a hand-pump mounted in a barrel or tank and equipped with two leads of 3/8 inch hose 25 feet long, each with a four-foot, extension made from ¼ inch gas pipe, and a double Vermorel nozzle. The barrel should be carried in an ordinary farm wagon. Three men do the work. One is expected to drive and pump, while the other two manipulate the nozzles. The outfit is stopped while the plants within reach are sprayed, then driven forward about 30 feet and stopped again. On an average in actual field practice 3 to 4 acres a day can be sprayed in this way, applying 100 to 200 gallons of Bordeaux per acre. To keep the long hose off the plants two poles about 10 feet long may be pivoted to the bed of the wagon so as to swing at an angle over the wheel and carry the hose. The pump for this outfit should be of good capacity, with brass valves. A "Y" shut-off discharge connection on the pump is a convenience for stopping the spray at any time. The most satisfactory nozzles are those of the Vermorel type. It is cheapest in the long run to buy the best grades of pumps on the market. This outfit is excellently adapted for spraying small fields of potatoes and for general orchard work, and is invaluable on the average farm.

Phytoptosis.—This disease is known to occur only in Florida, where it is sometimes common enough to require remedial treatment. The affected portions of the foliage are more or less distorted and covered with an ashy white fuzz. The general vigor and fruitfulness of the plants are greatly reduced. The name [Pg 139] applied to this trouble denotes its cause, an extremely small mite (*Phytoptus calacladophora* Nal.), which by its presence on the leaves or stems so irritates them as to result in the abundant development of modified plant hairs, which shelter the mites and form the fuzzy covering characteristic of the disease. A remedy for phytoptosis is available in the sulphur compounds. The following one is particularly recommended by Prof. P. H. Rolfs, to whom our knowledge of the disease is due:

Preparation of sulphur spray.—Place 30 pounds of flowers of sulphur in a wooden tub large enough to hold 25 gallons. Wet the sulphur with 3 gallons of water, stir it to form a paste. Then add 20 pounds of 98 per cent. caustic soda (28 pounds should be used if the caustic soda is 70 per cent.) and mix it with the sulphur paste. In a few minutes it becomes very hot, turns brown, and becomes a liquid. Stir thoroughly and add enough water to make 20 gallons. Pour off from the sediment and keep the liquid as a stock solution in a tight barrel or keg. Of this solution use 4 quarts to 50 gallons of water. Apply with a spray pump whenever the disease appears, and repeat if required by its later reappearance. The use of dry sulphur is also recommended.

DISEASES OF THE FRUIT

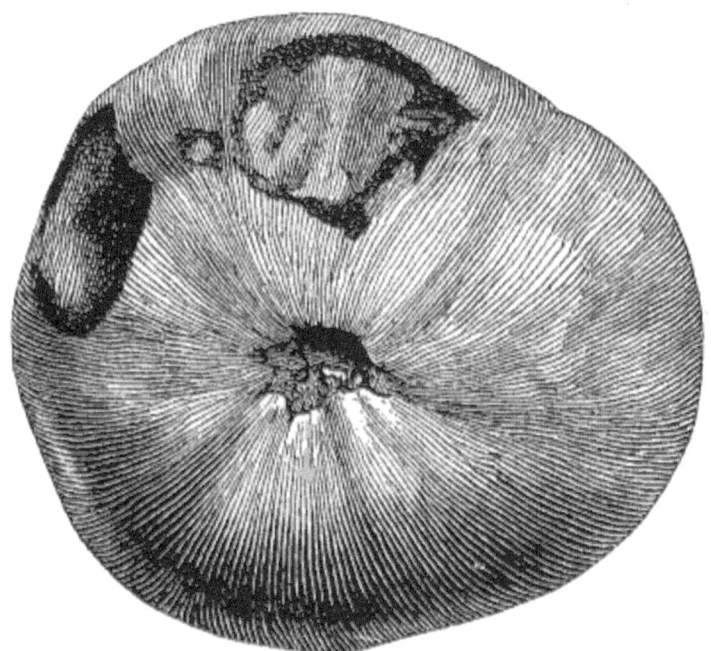

FIG. 43 — POINT-ROT DISEASE OF THE TOMATO
(Redrawn from N. Y. Expr. Sta. No. 125)

Point-rot. — This trouble, called also "blossom-end rot," and "black-rot," occurs on the green fruit at various stages of development, as shown in Fig. 43. It begins at the blossom end as a sunken brown spot, [Pg 140] which gradually enlarges until the fruit is rendered worthless. The decayed spot is often covered in its later stages by a dense black fungous growth (*Alternaria fasciculata* (C. & E.) J. & G. syn. *Macrosporium tomato* Cke.), formerly thought to be the cause of the rot, but now known to be merely a saprophyte. Point-rot sometimes occurs in greenhouses, but is more common in field culture. It is one of the most destructive diseases of the tomato, but its nature is not fully worked out, and a uniformly successful treatment is unknown. It has been thought to be due to bacterial invasion, but complete demonstrations of that fact [Pg 141] have not yet been published. The physiological conditions of the plant appear to be important. The disease is worst in dry weather and light soils, where the moisture supply is insufficient, and irrigation is beneficial in such cases. Spraying does not control point-rot so far as present evidence goes.

Anthracnose — ripe-rot — (*Colletotrichum phomoides* (Sacc.) Chest.), is distinguished from the point-rot by the fact that it occurs mainly on ripe or nearly ripe fruits, producing a soft and rapid decay. Widespread losses from this cause are not common, but when a field becomes infected a considerable proportion of the crop within a limited area may be destroyed if humid or rainy weather prevails. Preventive measures only can be employed. These should consist in collecting and destroying diseased fruit and in staking and trimming the vines to admit light and air to dry out the foliage. Bordeaux mixture applied after the development of the disease would be of doubtful efficiency and would be objectionable on account of the sediment left on the ripe fruit.

DISEASES OF THE ROOT OR STEM

Damping off. — Young plants in seed-beds often perish suddenly from a rot of the stem at the surface of the ground. This occurs as a rule in dull, cloudy weather among plants kept at too high a temperature, crowded too closely in the beds or not sufficiently ventilated. Several kinds of fungi are capable of causing damping off, under such conditions.

Preventive measures are of the first importance. [Pg 142] Since old soil is often full of fungous spores left by previous crops, it is the wisest plan to use sterilized soil for the seed-bed. When the young plants are growing, constant watchfulness is required to avoid conditions that will weaken the seedlings and favor the damping off fungi.

Watering and ventilation are the two points that require especial skill. Watering should be done at midday, to allow the beds to drain before night, and only enough water for the thorough moistening of the soil should be applied. Ventilation should be given every warm day as the temperature and sunshine will permit, but the plants must be protected from rain and cold winds. Work the surface of the soil to permit aeration and do not crowd the plants too closely in the beds. If damping off develops something can be done to check it by scattering a layer of dry, warm sand over the surface, and by spraying the bed thoroughly with weak Bordeaux or by applying dry sulphur and air-slaked lime.

Bacterial wilt (*Bacterium solanacearum* Erw. Sm.).—This disease, which also attacks potatoes and eggplants and some related weeds, is one of the most serious enemies of the tomato. It is known to occur from Connecticut southward to Florida and westward to Colorado, but is most prevalent in the Gulf States, where it has greatly discouraged many growers.

Its most prominent symptoms are the wilting of the foliage and a browning of the wood inside the recently wilted stems. An affected plant wilts first at the top, or a single branch wilts, but later the entire plant yellows, wilts and dies. Young plants wilt more sud [Pg 143] denly and dry up. The disease progresses more rapidly in plants that have made a succulent, luxurious growth, while those with hard, woody stems resist it somewhat.

The disease is due to the invasion of bacteria, which enter the leaves through the aid of leaf-eating insects, or through the roots. They plug the water-carrying vessels of the stem, shutting off the water and food supply of the plant. If the stem of a plant freshly wilted from this disease be severed, the bacteria will ooze out in dirty white drops on the cut surface.

Remedial measures entirely satisfactory for the control of bacterial wilt have not yet been worked out. The best methods to adopt at present are the following:

(1) *Rotation of crops.*—The field evidence is that this disease is in many cases localized in old gardens or in definite spots in the field. It appears also that the infection left by a diseased crop can remain in the soil for some time. It is therefore advised that tomato growers should always practice a rotation of crops, whether any disease has appeared or not, and that in case bacterial wilt develops they should not plant that land in tomatoes, potatoes, or eggplants for three or four years. The length of rotation necessary to free the soil is not known, but will have to be worked out by the individual grower.

(2) *Destruction of diseased plants.*—The bacteria causing wilt not only spread through the soil but are carried by insects from freshly wilted to healthy plants. Diseased plants thus become dangerous sources of infection, and it is evident that all such should be pulled out and burned. This is particularly important at the [Pg 144] beginning of the trouble when the eradication of a few wilting plants may save the remainder.

(3) *Control of insects.*—To lessen the danger from spread of wilt by insects, the measures advised in the next chapter for the control of leaf-eating insects should be adopted. In this connection it should be mentioned that the use of Bordeaux mixture for leaf blights, as previously recommended, has an additional value in that the coating on the leaves is distasteful to insects and helps to keep them away.

(4) *Seed selection.*—Work done at the Florida experiment station indicates that resistant varieties may be secured, but there are as yet none in commercial use. This is an important line for experimenters to follow up. There is no proof that the disease is spread through seed from diseased plants.

Fusarium wilt.—This disease and the one following resemble the bacterial wilt so closely, as far as external characters go, that they are difficult to tell apart. The parasites, however, differ so materially in their nature and life history that the field treatment is quite different. There are also differences in geographical distribution that are important, for while the Fusarium wilt occurs occasionally throughout the southern states, it is known to be of general commercial importance only in southern Florida and southern California.

The symptoms of the disease are a gradual wilting and dying of the plants, usually in the later stages of their development. Young plants die, however, when the soil infection is severe. There is a browning of the woody portions of the stem, and when a section of this is examined under a compound microscope the [Pg 145] vessels are found to be filled with fungous threads, which shut off the water supply.

The infection in the Fusarium wilt appears to come entirely from the soil. Little is known of its manner of spread, except that the cultivation of a tomato crop in certain districts appears to leave the soil infected so that a crop planted the next year will be injured or destroyed. The fungus does not remain in the soil for a very long time in sufficient abundance to cause serious harm. A rotation of crops that will bring tomatoes on the land once in three years has been found in Florida to prevent loss from Fusarium wilt.

This fungus does not attack any other crop than tomatoes, so far as known, though it is very closely related to species of Fusarium producing similar diseases in cotton, melon, cowpea, flax, etc. Fusarium wilt has not been found in fields and gardens in the northern states, but tomatoes in greenhouses there are sometimes attacked by it or a related Fusarium, which also occurs in England. When greenhouse beds are infected the soil for the next crop should be thoroughly sterilized by steam under pressure.

Sclerotium wilt.—This disease resembles the two preceding in its effect on the plant, which wilts at the tip first, and gradually dies. Its geographical range is more restricted. It seems to be confined to northern Florida and the southern part of Georgia and Alabama, where it occurs in gardens and old cultivated fields. The fungus causing this wilt attacks the root and the stem near the ground, working in from the outside. There is not the browning of the wood vessels characteristic of the two preceding diseases. If an [Pg 146] affected stem is put in a moist chamber made from a covered or inverted dish, there will develop an exceedingly vigorous growth of snow-white fungous mycelium which, after a few days, bears numerous round shot-like bodies, at first light-colored, then becoming smaller and dark-brown. These are the sclerotia or resting bodies of the fungus. This fungus, called *Sclerotium* sp., or "Rolf's Sclerotium," is noteworthy because it attacks potatoes, squash, cowpea, and a long list of other garden vegetables and ornamental plants. The only satisfactory means of control is rotation of crops, using corn, small grain, and the Iron cowpea, a variety immune to this and other diseases. Susceptible crops should be kept from infected fields for two or three years.

Root-knot (*Heterodera radicicola* (Greef) Mül.) attacks tomatoes in greenhouses and is in some cases an important factor in southern field culture. It is caused by a parasitic eelworm or nematode, of minute size, which penetrates the roots and induces the formation of numerous irregular swellings or galls, in which are bred great numbers of young worms. The effect on the plant is to check growth and diminish fruitfulness, in advanced cases even resulting in death.

The remedy in greenhouse culture is thorough soil sterilization. In the open field this is impracticable and recourse must be had to a rotation with immune crops, which will starve out the root-knot. It must now be borne in mind that the root-knot worm can attack cotton, cowpea, okra, melons and a very large number of other plants. The only common crops safe to use in such a rotation in the South are corn, oats, velvet [Pg 147] beans, beggar weed, peanuts, and the Iron cowpea. The use of other varieties of cowpea than the Iron is particularly to be avoided, on account of the danger of stocking the land with root-knot. Fortunately, the disease is serious only in sandy or light soils.

Rosette (*Corticium vagum* (B. & C.) var. *solani* Burt.) is a disease of minor importance, which occurs in Ohio, Michigan, and scatteringly in other states. The fungus causing it (*Rhizoctonia*) attacks the roots and base of the stem, forming dark cankers. The effect on the plant is to dwarf and curl the leaves and to restrict productiveness. The potato suffers more severely from the same trouble. Rotation of crops and liberal application of lime to the soil are advised for the control of rosette in tomatoes.

[Pg 148]

INDEX

PAGE

Adaptations of varieties, 97
as to habit, 97
as to foliage, 100
as to fruit, 102

Botany, 1

Canning, cost of, 118
on the farm, 118
Essentials for successful, 119

Catalog descriptions incomplete, 110

Characteristics of blossom, 25

Characteristics of fruit, 26
Development from original form, 26
Effect of conditions on, 26
Quality, 26

Characteristics of plant, 20
Checking of growth, effect upon, 20
Natural environment, 20
Uniform growth, importance of, 21

Characteristics of root, 23

Characteristics of stem and leaves, 24

Classification, 4
Cherry, 5
Cultivated varieties, 10
Currant, 4

Pear, 7

Cold-frames, construction, 53

Commercial importance of crop, 18

Cost of crop, per acre, 121
as grown for canners, 117

Covers for plant beds, 55

Cultivation, 76

Care and thoroughness necessary, 76
in greenhouse, 77
in home garden, 77

Diseases, 131
Bacterial wilt, 142
Blight, early, 135
Blight, leaf, 134
Blight, Western, 134
Cracking of fruit, 132
Damping off, 141
Edema, 133
Fusarium wilt, 144
Leaf curl, 132
Leaf mold, 135
Leaf spot, 134
Mildew, downy, 135
Mosaic disease, 133
Phytoptosis, 138
Point rot, 139
Root knot, 146
Sclerotium wilt, 145
Yellows, 134

Diseases, remedies for, 131

Bordeaux mixture, preparation of, 136
Preventatives of, 143
Spraying apparatus, 137
Spraying, importance of, 136
Sulphur spraying, 139

Distances for setting plants, 68
in field, 68
in greenhouse, 70
in home garden, 69

Drainage, importance of, 31

Essentials for best development, 28
Cultivation, 32, 76
Effect of shade, 28
Food supply, 31, 43
Heat, 30
Moisture, 30
[Pg 149] Sunlight, 28

Exposure, 38
for early crop, 39
for greenhouse, 40
for home garden, 40

Fertilizers, 43
Amounts, 43
Character, 44
Experiments with, 45
for general application, 44
for greenhouse, 45
for home garden, 45

Flats, construction, 57

Gathering fruit, 91

Habit, 22

Handling fruit, 92

History, 14

Hotbeds, construction, 51

Hotbeds, growing fruit in, 70

House, construction, 49

Insects injurious to tomatoes, 123
Blister beetle, 125
Colorado potato beetle, 125
Cut worm, 123
Flea-beetle, 124
Stalk-borer, 127
Tomato fruit worm, 128
Tomato worm, 126
White fly, 130

Location of field as determining profit, 38

Manure
Fall dressing, 41
for cold-frames, 55
for greenhouse soil, 37
for hotbeds, 51
in preparing ground, 46

Origin, 10

Origin of name, 14

Packing, 94

Pollinating, 77

Pollination, 25

Prices obtained
at canneries, 118
for hothouse fruit, 122
for select field grown fruit, 122

Profits on crop, 122

Propagation of plants, 48
from cuttings, 49
from seed, 48, 49
in cold-frames, 53
in hotbeds, 51
in temporary greenhouses, 49

Pruning, 80

Ripening on the vines, 90

Ripening after frost, 95

Sash, cost, 49
for hotbeds, 52
for cold-frames, 53

Seed breeding, 112
Essentials to success, 113
Growing and saving commercial seed, 115

Methods followed,   115
Prices received,   116
Yields obtained,   116
Importance of breeding from individual plants,   114
Importance of exact ideals,   113
Methods recommended,   113
Principles underlying,   112

Setting plants,   70
Conditions favorable and unfavorable,   70, 71
in field,   70
in greenhouse,   74
in home garden,   74
New Jersey method,   71
Other methods,   73

Soil
Composition, importance of,   24
Conditions essential,   41
Preparation,   41, 46
for greenhouse,   47
[Pg 150] for home garden,   47

Soil Preparation,
for main crop,   46
Importance of,   46
Selection,   33
for early crop,   36
for greenhouse,   37
for home garden,   36
for main crop,   34
Previous crop,   41

Sorting,   92

Staking,   79

Starting plants,   59
Effect of shade,   29

for early fruit,   63
for forcing,   67
for home garden,   67
for late crop,   65
in flats,   59
in greenhouse,   59
Pricking out,   60
Right conditions,   62
Spotting boards,   61
Unfavorable conditions,   63
Watering,   60
With least labor,   66

Succession, practice in the South,   42

Training,   79
for greenhouse,   88
for home garden,   85

Types,   14

Value, development of,   16

Variations,
in foliage,   100
in fruit,   102
Coloring,   106
Flesh,   105
Ripening,   106
Shape,   102
in habit, 97

Varietal differences,
as to foliage,   100
as to fruit,   102
as to growth,   97

Variety names, 108
Sources, 109
Varying application, 110

Watering, danger in, 30, 60

Yielding capacity, 22

Yield per acre, 117, 121

Yield per foot of greenhouse bench, 122

www.ingramcontent.com/pod-product-compliance
Lightning Source LLC
Chambersburg PA
CBHW031422210526
45464CB00005B/2006